Advance Praise

"*Out-think!* is a lucidly written book on introduction to game theory and strategic thinking. I'm sure, decision-makers and functional managers would find *Out-think!* really useful. Moreover, it will certainly interest final year undergraduate and postgraduate students of management, economics, politics and engineering."
Sanjay Singh, *Professor, IIM Lucknow*

"This book introduces to the reader, a student or a practitioner, how key strategic business decisions—Pricing, Market Entry, Marketing, Recruitment, Negotiation, etc.—have underlying principles based on Game Theory. One dedicated course does not do justice to the principles covered. The book should be recommended as a reference for various business courses because of the case driven approach. As a practitioner the book acts as a refresher to the various principles taught by Dr Sarkar, that I continue to use in my professional life."
Abhishek Kumar, *Principal Consultant, Utilities Leader* **E&E, APAC, Frost & Sullivan**

"A wonderful mental workout for every strategist, game theory professor and aspiring CEO. Full of paradoxes, puzzles and flashes of insights. I am sure this book can cross over from being used in the classroom to keeping the mind ticking in the airport lounge."
Dr Rohit Prasad, *Associate Professor, MDI Gurgaon*

OUT-THINK!

OUT-THINK!

How to Use Game Theory
to Outsmart Your Competition

SUMIT SARKAR

www.sagepublications.com
Los Angeles • London • New Delhi • Singapore • Washington DC

First published in 2016 by

SAGE Response
B1/I-1 Mohan Cooperative Industrial Area
Mathura Road, New Delhi 110 044, India

SAGE Publications Inc
2455 Teller Road
Thousand Oaks, California 91320, USA

SAGE Publications Ltd
1 Oliver's Yard, 55 City Road
London EC1Y 1SP, United Kingdom

SAGE Publications Asia-Pacific Pte Ltd
3 Church Street
#10-04 Samsung Hub
Singapore 049483

Published by Vivek Mehra for SAGE Publications India Pvt Ltd, typeset in 11/13 pt Bembo by Diligent Typesetter, Delhi and printed at Sai Print-o-Pack, New Delhi.

Library of Congress Cataloging-in-Publication Data

Sarkar, Sumit (Economist), author.
 Out-think! : how to use game theory to outsmart your competition / Sumit Sarkar.
  pages cm
Includes bibliographical references and index.
1. Game theory. 2. Decision making. 3. Strategic planning. I. Title.
HD30.26.S26 658.4'033—dc23 2016 2015032507

ISBN: 978-93-515-0563-1 (PB)

The SAGE Team: Sachin Sharma, Guneet Kaur Gulati, Apeksha Sharma, Nand Kumar Jha, and Ritu Chopra

*To Arundhati and Ashmani, without whose support
it would not have been possible*

Contents

List of Case Studies ix
Preface xi
Acknowledgements xvii

1. What Managers Can Learn from Game Theory? 1

2. Business and Chess: Looking Forward, Reasoning
   Backward 9

3. Prisoner's Dilemma 31

4. Coordination and Anti-coordination Games 48

5. Strategic Moves: Threats, Promises and Commitment 70

6. Trust, Credibility and Collusion in Repeated Games 89

7. Business Poker: Playing Games with Limited
   Information 115

8. Smart Negotiations 164

Epilogue 201
Bibliography 206
Index 210
About the Author 214

List of Case Studies

2.1   Starbucks' Entry in Indian Café Retail   18
2.2   Boeing, Airbus and the Ultra-high-capacity Airliner   28

3.1   Battleground Iberia—Boeing versus Airbus   37
3.2   Tobacco Advertising on TV—USA (1950–70)   43

4.1   RFID Tagging at Metro AG   56
4.2   War of Attrition in Satellite Television Market of UK   65

5.1   Hindustan Lever Ltd and Nirma   82
5.2   Starbucks and Costa in China   86

6.1   Crude Pricing by OPEC   103
6.2   Price Fixing by DuPont and Its Competitors   108
6.3   Price Match Guarantee by Toys"R"Us   113

7.1   Predatory Pricing and Walmart   137
7.2   Ola's Alleged Predatory Pricing   143

8.1   Individual Behaviour in Simple Bargaining Games—
An Experiment in a High Stake Ultimatum Game
Conducted in Villages of Meghalaya   172
8.2   A Teaching Case in Alternate Offer Bargaining—
Acme Wagon Co. versus Selco Steel Inc.   179

Preface

More than 70 years ago, in 1944, John von Neumann and Oskar Morgenstern published their book *Theory of Games and Economic Behavior*, wherein strategic decision-making scenarios were first defined as 'games'. Pioneering work done by John C. Harsanyi, Johan F. Nash and Reinhard Selten during the decades of 1950s and 1960s, in the development of the theory of non-cooperative games, got them 'The Sveriges Riksbank Prize in Economic Sciences in Memory of Alfred Nobel' in 1994. By then game theory had become a dominant analytical tool for economists. The fact that in the last 20 years between 1994 and 2014 seven Nobel prizes in Economic Sciences were shared between 18 scholars who either furthered the advancement of game theory or applied game theory in analysing economic problems stands as a testimony of importance of game theory in the discipline of economics. The 1994 Nobel Prize in Economic Sciences popularized game theory among academia outside the domain of economic theory and forced a section of the business academia to reconsider it as an applicable tool for management research. Till then, game theory was considered as a highly mathematical tool, useful only for development of 'esoteric' economic theory. During the two decades following the 1994 Nobel Prize, numerous research papers were published in journals, such as *Marketing Science, Decision Analysis, Management Science, Operations Research, Journal of Finance, Strategic Management Journal, Organizational Behaviour and Human Decision Processes, Group Decision and Negotiation*, etc., which applied game theory in solving problems in functional management areas of marketing, finance, operations and human resources, apart from the area of strategic management. During these 20 years, applicability of game theory in solving management problems was noticed by consultants such as Mackenzie & Company and Boston Consulting Group. At the same time, MBA curricula across the world incorporated game theory in their syllabi. This dual effect induced the strategic thinkers, corporate and business planners and consultants to accept game theory as an important management tool. Functional managers who engage themselves in negotiating and contracting with customers or suppliers, those who participate in bidding for contracts against competitors and those who are engaged in strategic and tactical decision-making—say in pricing—have, over time, realized the relevance of game theory to their profession. The number of executive development programmes on functional application of game theory, conducted by the schools of businesses in the universities of both developed and emerging economics, also proves that game theory has become an integral part of the practicing manager's tool box.

I grew up as an academician during this very period—1994 to present. After finishing my doctoral research in the field of industrial organization, a discipline that relies heavily on application of game theory, I have taught game theory to students of management, and to students of engineering, since 2004, in various premier institutions of higher learning in the country. Teaching game theory to MBA students enriched my understanding of the subject. It is easy to catch the imagination of bright students with the analytical sharpness of the subject. But, at the same time, MBA students always ask for 'real-life examples' to see how the analytical rigour is useful in managing day-to-day businesses. In order to ensure respectable feedback from my students, I had to dig out real-life business cases and show them how the situation could be analysed using game theory tools. These 11 years of interaction with bright, young minds in the MBA classrooms has been a huge learning experience for me. This book is essentially an output of that learning experience, founded upon the base that was made by my teachers and thesis supervisor during my student days.

Conflict along with strategic nature of relations breeds scenarios that are games. Tactical game play involves minute observation of the game situation, identifying weaknesses of the rival(s) and exploiting those weaknesses to outsmart them. Games such as football or chess, and situations during armed conflicts are areas where tactical game play is of utmost importance. Business had been seen as conflict scenarios and often compared with war. In fact, that is a bit of a cliché now. New age management gurus talk about sustainability, cooperation, blue-ocean strategy and spirituality. In this era of these new-found areas of wisdom, game theory runs the risk of being branded as a tool with narrow and myopic scope. This misconception comes from the idea that game theory begins and ends with an apparently indigestible game called prisoner's dilemma. One purpose of this book is to clear that misconception. Indeed, it is a tool for tactical game play in competitive scenarios, but it also helps in identifying win–win situations and opportunities to cooperate. In fact, that very example of prisoner's dilemma can be used to show how it is possible for rivals to collude and increase payoffs. Adam Brandenburger and Barry Nalebuff, in their 1997 book *Co-opetition*, nicely established the need for businesses to combine cooperation with competition.

It is possible to analyse a wide variety of business scenarios applying game theoretic techniques. In this book, I used case studies from different industries to focus on issues such as pricing, market entry, technology adoption, new product development, negotiations, bidding, etc. Competition is inherent in most of these scenarios, and throughout this book I maintained the fundamental assumption of game theory that game players are self-interested. This assumption, per se, is not often contested as not many people suffer from the misconception that businesses are done with an altruistic motive. But the assumption that players can make perfect rational decisions is often contested. A serious chunk of the criticism against the assumption of rationality comes from the practicing managers, for whom I have been conducting training sessions covering modules on competitive strategy, pricing, positioning, etc., for over little more than a decade. In these modules, I show how game theory can be used in making strategic decisions. The participants come from middle to senior management ranks of various organizations, which include the public sector companies as well as private corporations. Scepticism with the assumption of rationality is across the board. Interestingly, the practitioners who contest the assumption of rationality think that they themselves are capable of making decisions in the most rational fashion without getting carried away by emotions, but they are not sure whether their suppliers, customers and competitors are capable of the same. If most people think that they can take decisions rationally, and they actually can, then there is little chance for the devil called irrationality. Even if there is a positive chance that your rival is irrational, game theory can deal with such situations by considering it as a game of incomplete information. In this book, I will address games of incomplete information in Chapter 7. Sometimes what seems to be irrational may have a different sort of rationale that doesn't occur to the naive. I will address such apparent irrationality in Chapter 5.

Nevertheless, it is true that many game theoretic arguments degenerate if the assumption of rationality is violated at the fundamental level. There are two sources of irrationality in decision-making—lack of cognitive ability and emotions affecting decision-making. Individual decision-makers suffer from both. There is vast literature on evolutionary game theory that does not presume individual game players to be born rational. This literature argues that individuals are not rational, but they try to be rational. As they get used to a particular environment, they become more and more capable of making rational decisions. Emotions, on the other hand, can be addressed within the framework of rational decision-making by attaching payoffs on intangible aspects such as pleasure, revenge, etc. Behavioural game theory deals with emotions alongside rationality. In this book, I won't cover evolutionary game theory, or behavioural game theory. The scope of this book is limited to the applications of game theory in situations where players are capable of making rational decisions. Business decisions, particularly in the business-to-business (B2B) context, fit the bill. Strategic business decisions are normally made by a think tank comprising of a group of qualified individuals. These individuals might act only as bounded rational agents if they were making decisions individually. But when they brainstorm in a group, the learning is much faster, and the group should be capable of making rational choices. When decisions are made by a group, the role of individual emotions can be minimized unless some members of the group are powerful enough to override others. Despite businesses' ability to make rational decisions, game theory may not work in a business-to-consumer (B2C) set-up as the consumers might not be capable of making rational choices. Keeping that in mind we will focus on the use of game theory in decision-making in B2B set-ups.

The issue of ethics is of utmost importance in modern businesses. There is a huge misconception that knowledge of game theory makes decision-makers unethical. However, half-baked knowledge of game theory may make decision-makers act unethically. Also, only a person who is half educated in game theory thinks that game theory and ethics can never go hand in hand. The source of this misunderstanding is again a muddled up idea of individual rationality. In my opinion, ethics is long-term rationality. Experts in business ethics will possibly disagree. But they disagree amongst themselves on what is ethical. Is a marketer of a tobacco product unethical? Some will say no as the marketer is doing it for a living, and he/she is being ethical to his/her employer. Even a peddler of soft drugs like marijuana does it for a living. Why not legalize marijuana then? On the other hand, if selling tobacco products is unethical, then why not ban tobacco? Game theory does not address questions like these, but it also does not teach you to cheat. Games such as football or chess are governed by a set of rules. Laws define the rules of business games. If there exist loopholes in the rules of a game, that is, if the rules are not well defined in all regards, self-interested players will want to manipulate the rules to their own advantage. Game theory is a handy tool in designing mechanisms to prevent such opportunistic behaviour. There is a large body of literature in the knowledge domain at the interface of law and economics. Game theoretic mechanism designing is indispensable in the area of law and economics. In this book, I will not cover mechanism designing as that is not of primary interest to businesses.

Decisions based on game theoretic analysis sometimes critically hinges on the payoffs. The question is: How do we get the payoff figures? There is a simple exercise in Chapter 2 that will give you an idea of how payoffs can be arrived at. But in the rest of the book I will take payoffs as given. Predicting the payoffs is not within the scope of game theory. Game theorists take the payoffs as given and analyse the situation with those given payoffs to arrive at a decision. In order to predict the payoffs, one needs advanced tools in analytics including data mining and forecasting.

Though it is rare, sometimes I encounter situations in the classroom, particularly in training sessions, where the participants expect game theory to be a handy tool wherein decisions will be generated if data is provided as input to a computer program. It is possible to develop computer programs that will do that for a given situation. But the purpose of learning game theory is to get hold of the logic. If the logic is clear, one will be able to construct his/her decision-making situation as a game and find out the optimal decision. Contrary to the popular belief, it is actually a very flexible tool. At the core, game theory is a way of thinking. I always tried to impart that way of thinking to my students. This book too attempts the same with its readers.

Acknowledgements

Since I am not Robinson Crusoe, I am indebted to many individuals for my very existence. Hence, I am indebted to an uncountably finite number of individuals who contributed indirectly to the preparation of this manuscript. I apologize for my inability to mention each one of them separately here. This book is the output of a decade's experience of teaching game theory to MBA students and practicing managers. I must at least mention all those who contributed to my professional development in the field.

I would have not developed any interest in game theory, or might have not even known what it is, but for two outstanding scholars—Professor Krishnendu Ghosh Dastidar and Professor Kunal Sengupta, both of who were my teachers at my alma mater Centre for Economic Studies and Planning, JNU, New Delhi. My training in game theory laid the foundation, but I would have not been able to write this book had I not been inducted into the faculty of IIM Kozhikode (IIMK) immediately after the completion of my doctoral studies. At IIMK, I got my first opportunity to teach game theory in the post graduate programme (PGP) class. I am thankful to the institute as well as my colleagues at IIMK, Professor P. Balakrishnan and Professor Nandakumar, for giving me the opportunity. In the years to follow, I got opportunities to teach the subject at IIT Kanpur (IITK), IIM Indore, IIM Ranchi and XLRI Jamshedpur. I take this opportunity to thank all these institutes and my colleagues therein. In particular, I would like to thank Professor Surajit Sinha and Dr Sanjay Singh, who were my colleagues at IITK, Professor Dipayan Dutta Chaudhury of IIM Indore and Professor Suma Damodaran of XLRI Jamshedpur. The above-mentioned institutions provided me the opportunities, but my students at these institutions enriched my experience. I thank each of them for their thought provoking questions and lively discussions within and outside of the classrooms. PGP or MBA students provoked me to think to answer them and to make game theory more applicable in functional disciplines of management. But the participants in training sessions brought in their experiences to the classrooms and enriched my understanding. I am thankful to each of the participants who attended my training sessions. I take this opportunity to thank Professor Krishna Kumar, then Director of IIMK, for providing me the first opportunity to teach in a training session for senior managers of a public sector unit. I also thank my colleagues at XLRI, Professor Sabyasachi Sengupta, Professor N. Rajkumar, Professor D. P. Sinha, Professor P. K. Padhi and Professor Santanu Sarkar for giving me opportunities to take sessions on varying topics including pricing, bargaining, competitive strategies in their respective management development programmes for various organizations. Special thanks to Professor M. G. Jomon for ensuring that my management development programme on game theory had enough participants. I thank Tata Chemicals, Mahindra Finance, NABARD, Dr Reddy's, Forbes Marshall Pvt. Ltd and various other organizations that sent their managers to attend my training programmes.

I thank SAGE Publications and Mr R. Chandra Sekhar for trusting me with this commission. Special thanks are reserved for Mr Sachin Sharma of SAGE Publications who helped me with the preparation of the manuscript. I possibly would have not completed this book but for Sachin. I acknowledge the efforts of the SAGE team comprising Guneet Kaur Gulati, Nand Kumar Jha and Apeksha Sharma in completion of this book. I must also thank Mr Premendra Sharma from SAGE Publications, who put me in touch with Mr Chandra Sekhar and Sachin.

My family was deprived of my time with them because of my preoccupation with writing this book. Ashmani had to accept her dad working on the computer even during the evenings and on weekends. I am grateful to Arundhati for running the household and sacrificing her own research projects. Discussions with her on the subject were always helpful. I also thank her for designing the cover of *Out-think!*

1

What Managers Can Learn from Game Theory?

*B*CG *Perspectives* published by Boston Consulting Group, dated 3rd December 2009, mentions a problem encountered by one of their clients. The nature of the problem is generic. The article states:

> One of our clients learned this the hard way. Convinced that the sales force was giving away too much in price negotiations in order to capture volume, this company undertook a pricing project in which it analysed accounts, identified opportunities to raise prices, and provided a new set of pricing guidelines. The resulting profit boost was quick and significant. Unfortunately, a short while later, the company found itself back in its original position and in need of another pricing remedy. The problem resurfaced because the leaders of the sales force continued to drive a culture that emphasized volume, rather than profitability. Without changing its incentives, processes, and people, the company could not achieve sustainable impact from pricing improvements.

The scenario is very familiar to most practising managers, strategic planners and consultants. The genesis of the problem seems to be organizational culture. Culture gets perpetrated by the game you make people play. In order to change the culture, it is imperative to change the game. Incentives are keystones of strategic game playing, and to change the game you need to change the incentive structure.

You possibly face similar problems in various spheres of your business. When you negotiate a contract with suppliers, or clients, or employees, you negotiate on the rules of the game. When you draw the rules, you need to make sure that others are incentivized to play the game as per the drawn rules. At times you are sucked into games where the rules are made by someone else—maybe by the market or the regulators. Irrespective of whether you chalk-out the games or you are forced to play it as per rules made by someone else, you play so you are. In Latin they say *ludo ergo sum*—I play so I am! They also say *cogito ergo sum*—I think so I am! To play you need to think the right way. This book will help you in shaping your way of thinking in contexts of games you might have to play.

This book attempts to shape up your strategic thinking with help of various examples from the field of business. Many strategic or tactical moves apparently seem puzzling to us. For example, Costa Coffee has been setting up coffee shops within a stone's throw of Starbucks outlets in China. To the naive it seems to be a stupid strategy. Why would Costa do it? What is the game? And if you are Starbucks, how do you react to this? Microsoft spends around US$13 billion per year on R&D and a large part of it is spent on tuning future versions of Windows and Office. You might wonder why they need to spend so much on upgrading products in which they are unchallenged category leaders. Is there some game behind it? What is the game, and who does Microsoft play the game against? This book will not only help you in comprehending the underlying games of such apparently puzzling strategic moves, but also teach you a few tricks of playing such games.

In order to do so we will use the techniques of game theory. But instead of developing complex mathematical theories we will develop strategic thinking by drawing upon examples from different walks of life including politics, international relations, geopolitics, military history and sports. The purpose of this book is to make game theory understandable and usable for strategic decision-makers and functional managers. Will that make you a better manager? It will help you to out-think and outmanoeuvre your competition, suppliers, complementors and employees, or at least to keep up in the tactical battles. But it is frills-free. It does not teach you how to maintain eye contact during negotiation, nor does it teach you about power-dressing to dominate over your rivals. To get going, let us first understand what game theory is.

What Is Game Theory?

Game theory is the science of analysing scenarios that can be described as games. A game scenario involves strategic interaction between two or more self-interested players, who are aware of their own gains and losses from different plausible outcomes of the strategic interaction. The description of a game is required to specify the following:

1. *Players*: A group of entities, individuals or organizations, are players of a game if they are involved in strategic interaction wherein their decisions affect each other's well-being or happiness. Game theory assumes that these players are self-interested and that they are aware of the fact that every game player is self-interested. The players are smart enough to process any information available to them. Emotion plays no role in their decision-making. At the onset let me clarify the assumption of self-interested behaviour, which is also known as 'individual rationality'. Self-interested behaviour does not rule out cooperation among players. However the players won't cooperate just because it is nice to cooperate, but because they gain from cooperating rather than from competing. In the world of game theory, there is no such thing as altruism. However, actions that are thought of as 'altruistic' can be explained within the paradigm of 'individual rationality'. Baseline assumption is that players don't act without self-interest, but the scope of self-interested behaviour stretches beyond the sphere of economics and politics. It might just be spiritual. It might even be a 'warm glow', which

is defined as a satisfaction from increasing the well-being of someone else. So, an action that appears to be altruistic must also involve some sort of self-interest.

2. *Strategies*: Strategies are actions or plans of action available to the players. The strategy set defines the scope of what the players can do in the game.

3. *Payoffs*: The description of a game must outline all the plausible outcomes of the strategic interaction. Payoffs are what the players get subject to realization of each of the outcomes. The payoffs of a player reflect the player's stake in the game. The assumption of self-interested players essentially means that the players strive to maximize their own payoffs.

Tales of Out-thinking Rivals

Game theory is interesting because there is tactical interaction. The players try to figure out their rivals' strategy before they move. Strategic games like chess or checker, football, basketball and such other team sports, battles on the warfront and international geopolitics are some spheres of life where everyone tries to out-think their rivals. Game theory can be used to analyse tactical moves in all these fields. Businesses can use game theory in the same manner to out-think competition. In this section we will walk through a few examples from different spheres of life to have an understanding of the phenomenon of out-thinking.

Out-thinking by Exploiting the Weakness of Rival

Spanish journalist Marti Perarnau, in his book *Pep Confidential,* quoted legendary football coach Pep Guardiola:

> *I sit down and watch two or three videos. I take notes. That's when that flash of inspiration comes—the moment that makes sense of my profession. The instant I know, for sure, that I've got it. I know how to win. It only lasts for about a minute, but it's the moment that my job becomes truly meaningful.* (Perarnau, 2014)

Guardiola was talking about 1 May 2009, the night before a crucial match for his team FC Barcelona against arch rival Real Madrid. The moment of magic was finding a new way to beat Real Madrid, using a then 21-year old Lionel Messi in a different role. Having watched a previous match between the two teams, Guardiola noticed a vast expanse of space between Real Madrid's midfielders and central defenders. That was a weakness asking to be exploited, and Guardiola took the opportunity. He asked Messi to move in to that space when Barcelona gained possession of the ball, and asked his midfielders to pass the ball to Messi. That move left the Real Madrid central defenders Cannavaro and Metzelde with two options. If they chase Messi they would leave the goalmouth exposed. A stroke of genius from Messi would see him dribble past the centre backs putting him one-on-one against the goalkeeper Casillas. On the other hand, if they hang back inside the penalty box they will be too late to go for the final tackle. Guordiola told Perarnau that he could visualize the situation on the night before the game, and summoned Messi at 10:30 PM to explain him his role. What Guardiola visualized got realized on the pitch 35 minutes into the game next evening. Messi's role in Barcelona was defined around this tactical masterstroke of Guardiola and Barcelona's strategy was built around that role. Yes, strategy was built upon around a tactical move.

Out-thinking by Doing the Unimaginable

During the last three weeks of the Second World War the 16th Armoured Division commissioned under the Third Army of USA captured the last few bases of Nazi Germany in Bavaria (southern Germany) and Czechoslovakia. One of the last strategic towns to fall was Pilsen, which is in the present Czech Republic. The town was strategic owing to the presence of Skoda munitions plant. An unconventional account of that military operation is narrated by Alan Cope, the protagonist of Emmanuel Guibert's graphic novel *Alan's War—The Memories of G.I. Alan Cope*. Alan Cope was a real character who served under General Patton in the Third Army of USA during the Second World War. On 6th May 1945 the

16th Armoured Division advanced along the Bor–Pilsen road. As per the account of Alan Cope, the division attacked Pilsen with only six tanks. They were backed by the rest of the division and troops from 2nd and 97th Infantry Divisions, but that back-up force was more than two hours behind the advanced tanks. The six advanced tanks fired heavily and enforced the Nazi forces to retreat. The Nazis didn't put up any resistance as they could not believe that there could be only six tanks. Once Pilsen fell, the Second World War was effectively over. The 16th Armoured Division effectively captured the town using only six tanks, which was incredible.

Out-thinking by Credibility

A fascinating story of out-thinking a hardball playing supplier is narrated by Adam Brandenburger and Barry Nalebuff in their book *Co-opetition*. Gainesville is a small town in Florida, USA. Gainesville Regional Utilities (GRU) supplies electricity to the town. For its captive power plant GRU used to buy coal from CSX which is one of the largest coal transporters in USA. They procured coal from the coal origins and delivered to Gainesville by railroad at a price of US$20.13 per ton in 1990. It was a monopoly price as no other railroad passed through Gainesville. GRU apparently got an upper hand over CSX when they got a deal from Norfolk Southern Railway (NSR) who agreed to deliver coal at US$13.68. The price was great for GRU, but NSR could not really deliver as their rail track did not pass through Gainesville. The NSR track passed through a junction with the CSX track 21 miles away. It was not feasible for GRU to take delivery 21 miles away and transport it on its own to Gainesville. So they asked CSX to let NSR coal trains use the CSX track to come to Gainesville. Of course CSX would have charged an access fee. But CSX refused. They didn't want to give up their monopoly position. GRU was unable to get out of the clutches of CSX who continued to overcharge them on coal. NSR was not ready to build their track for those 21 miles. So, GRU decided to construct their track for those 21 miles and let NSR use it to bring coal to Gainesville. The projected capital expenditure of building that track was US$28 million. If GRU got to buy coal at US$13.68 from NSR they save US$6.45 per ton. Simple calculation shows that they will recover their capital expenditure of US$28 million if they bought 4.35 million tons of coal from NSR at US$13.68. That is a huge quantity but numbers don't look absurd. In October 1991 CSX agreed to lower price by US$2.25 per ton, which was of course not a match for the NSR price. But the tactical game of out-thinking each other began from that point in time. CSX tried to outsmart GRU by threatening to abandon their track connecting Gainesville if GRU built their track. Indeed GRU didn't want that. If CSX abandoned their track and GRU built the track connecting Gainesville to the NSR track, NSR would become the monopoly supplier and GRU's bargaining power would be jeopardised in future. So, the threat of CSX was a credible one. But GRU didn't blink and went ahead with their plan. The proposed rail track was passing through a wetland and was awaiting clearance from the Environmental Protection Agency. There were hearings with the Interstate Commerce Commission too. NSR used its political influence to get the clearance from Environmental Protection Agency. In November 1992, when it was almost certain that NSR and Gainesville are going to get the clearance for building the track, CSX lowered price by further US$2.5 per ton. With US$5 reduction in price it didn't make economic sense to build the track any more. GRU abandoned the plan to build the track and signed a long-term contract up to 2020 with CSX. The contract resulted in a US$34 million savings for GRU in present discounted value (PDV) terms. With their credibility of being able to build the track they outsmarted CSX. But CSX wasn't the real loser in this case. It was Norfolk Southern.

Approach of the Book

Over ages academicians and thinkers have seen business scenarios as tactical battles. We see business scenarios as games. The business scenarios might be complex and it is impossible to out-think competition in such business games without a structured way of thinking. Game theory helps you in developing that structured way of thinking. The book is founded on the conviction that game theory, and only game theory, provides you the cutting edge in out-thinking your rival in competitive business situations. With that understanding the book attempts to walk you through the nuances of game theory.

Game theory is a complex tool and mathematics is used extensively in most of the formal game theory books. However, it is possible to learn the game theoretic way of thinking, and to apply that learning in decision-making, without using mathematics. In an attempt to make game theory understandable to almost everyone, this book keeps use of mathematics to a minimum. But the book contains many diagrams and figures, which are integral part of the analysis, and the readers will learn little if they skip those.

We will begin with simple game situations. The chapters are organized by concepts from game theory. The concepts are presented to the reader by means of simple examples instead of the conventional route of developing complex mathematical models. Every concept is put to use by applications in practical scenarios and real business cases. As we proceed, the game theoretic concepts become complex. The complexity is unavoidable as real-life scenarios are actually complex. This book hand-holds the reader and walks them through that complexity, and helps them in making decisions in such complex scenarios.

2

Business and Chess: Looking Forward, Reasoning Backward

I n a game of chess the players make moves sequentially. There are many business situations where the players move sequentially. For example, when Starbucks entered the Indian café scene in 2012, everyone expected them to take Café Coffee Day (CCD) head-on. That expectation was reinforced by the rate at which Starbucks started expanding in China since 2011–12. But Starbucks knows its games. Every game is different. A new country, a new rival, a new partner and it's a new game. CCD already owned more than 1,600 outlets and had presence in 200 cities and town across India. Instead of trying to challenge CCD with expansion, Starbucks targeted a different clientele and positioned itself as an upmarket player selling pricy coffee to executives in the business districts of Mumbai, Delhi and Bangalore. That was smart. But it was CCD who wanted to step on the gas. In a bid to have presence across the spectrum, they started opening upscale outlets called 'Lounge'. The Indian coffee war is brewing. Observe the moves. The moves are sequential like a game of chess.

Seeing through Your Rival's Strategy—
A Chess Example

A good chess player is able to see through the rival's strategy. While making his or her own move, a master chess player takes into account the plausible responses of the rival and reasons backward to make the move that is optimal. The following example will clarify the idea.

Suppose the white player makes an opening move by moving the king-pawn two ranks from e2 to e4 (Figure 2.1).

Figure 2.1

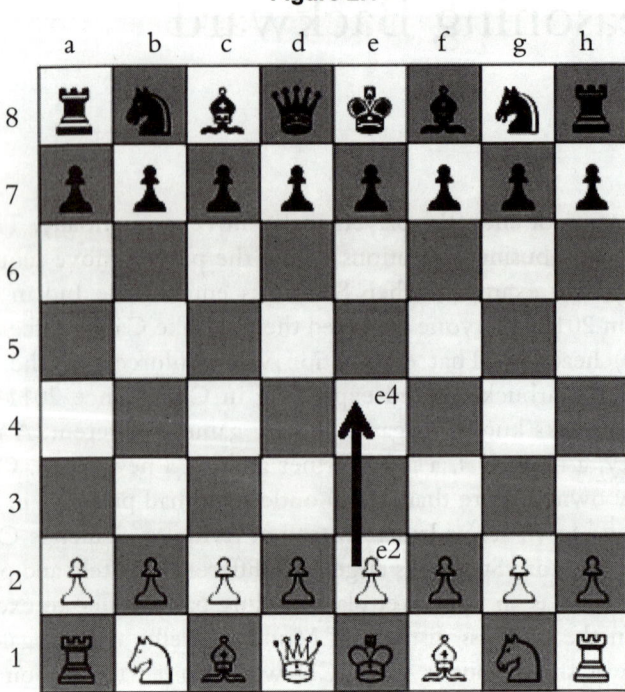

Before making any move, the black player should notice that the opening move of the white opened up a path for the white queen to move all the way to the edge on h5 (Figure 2.2).

Figure 2.2

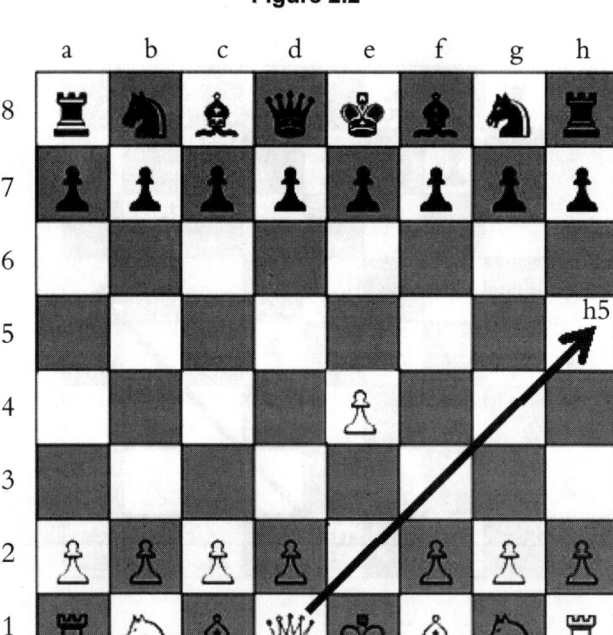

Black player should also notice that if the white player, in his or her second move, moves the white queen to h5, it puts the white queen on a diagonal vis-à-vis the black king, as shown by the dashed arrow in Figure 2.3. That is a potential threat of a checkmate and the only cover of the black king is the black bishoppawn in f7.

So, in his or her first move, the black player may keep the king protected by not moving the bishop-pawn from f7. But what should be the move? There are many possibilities. Black can block white's potential aggression by moving the knight-pawn one rank from g7 to g6 (Figure 2.4).

Now if the white moves his or her queen to h5, it will immediately be captured by the black pawn in g6, as shown in Figure 2.5. The box h5 is shaded in Figure 2.5 to indicate that it is a potential move, not an actual one.

Figure 2.3

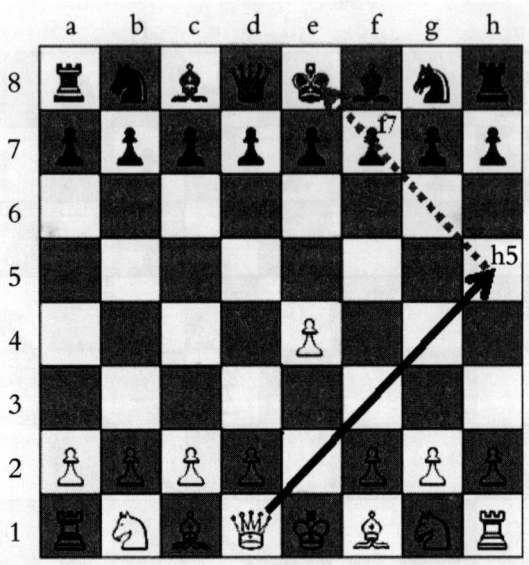

Figure 2.4

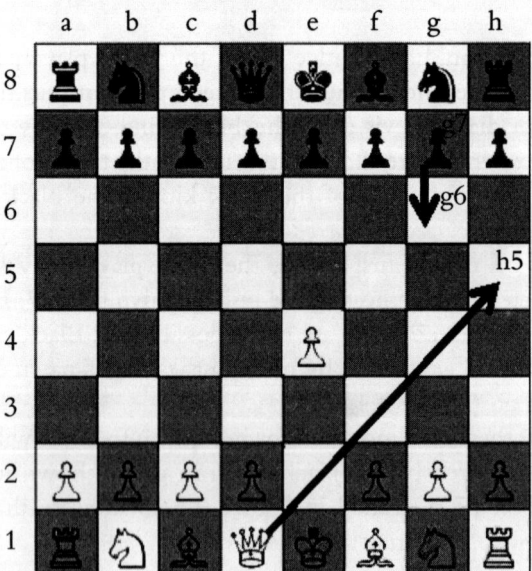

Figure 2.5

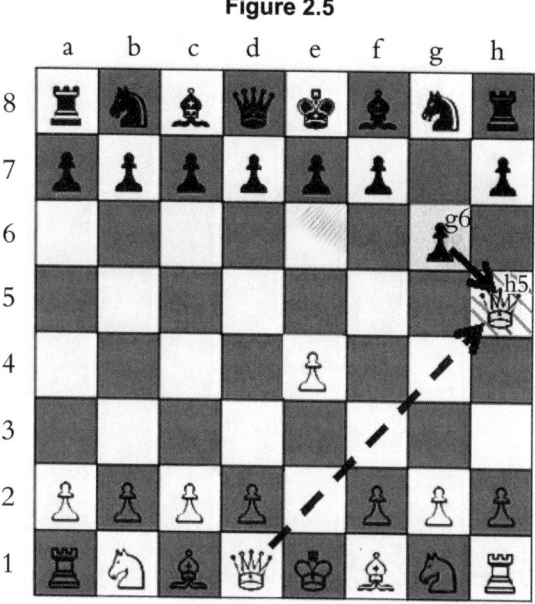

Anticipating the threat posed by the black pawn at g6 to his or her queen at h5, white player will not move the queen to h5. At most the white player can move the queen to g4 (shown in Figure 2.6), which does not threaten the defence of the black player.

Being able to foresee this effect, the black player moved his or her knight-pawn from g7 to g6, as shown in Figure 2.4.

In this example we strategized for black anticipating a threat from a potential movement of the white queen. But white's first move opened up a gully for moving the white bishop from f1 to b5, which movement will also potentially threaten the black king as it places the white bishop on the same diagonal with the black king. The defence of the black will be different if that threat is anticipated. This section is not meant to be a chess tutorial. Rather, the objective is to show you how to see thorough your rival's strategy. The first move of the black player in response to a particular opening move of the white and anticipating a particular attacking strategy of the white was shown as an example of looking forward and reasoning backward.

Figure 2.6

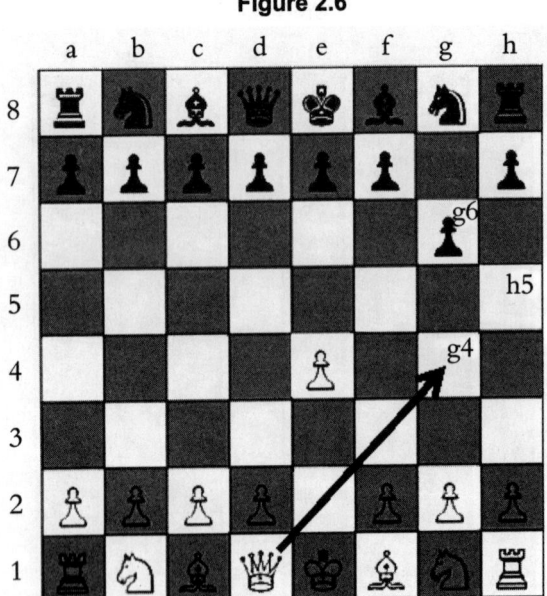

Looking Forward, Reasoning Backward— An Example

Chess is too complex a game to see through the entire game and reason backward to find the optimal moves of the players in each turn. Even the grand masters cannot see through the entire game. However, in relatively simpler games, it is possible to see through the entire game, anticipate each and every plausible move at each turn of each player, and reason backward to make correct decisions. To illustrate the idea of looking forward and reasoning backward, let us use a game called Chomp. This game was originally conceived by Dutch mathematician Frederik Schuh. American mathematician David Gale used the name Chomp in a particular formulation of the game.

The game is played on a 5 × 4 checker board with a ludo token placed at the bottom-left cell, as shown in Figure 2.7. Two players take turns to move the token on the board. As per the rule of the game, the token could be moved only one cell at a time.

Figure 2.7

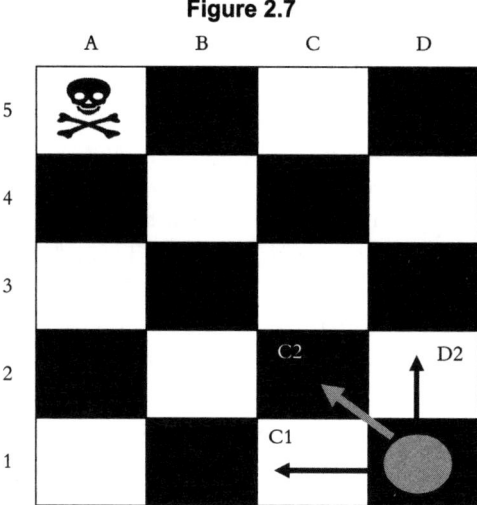

There are three valid moves. The token could either be moved one cell upwards, or one cell leftwards, or one cell diagonally up-left. Refer to Figure 2.7.

The cell A5 at the top-right corner is a marked as 'death' and is to be avoided. Whoever is forced to reach the 'death' cell loses the game.

At the beginning of the game the grey colour token is placed on the cell D1. Suppose you are the first mover. You may move the token either to D2, or to C1 or to C2. The second mover, who is your rival in the game, in his or her first turn will move the token one cell upwards, or one cell leftwards, or one cell diagonally up-left from wherever you left it in your first turn. That way the game continues till one of the players is forced by the other to move into cell A5, which is the 'death' cell. Suppose your rival is extremely intelligent and does not make any mistake. Or, you may suppose that you are playing against a computer program. In your first turn where will you move the token to, from D1? Will you move it to D2, or to C1 or to C2? In order to make the correct decision, you need to see through the entire game and reason backward.

You need to look forward all the way through the game, and visualize how the game might end. If your rival could take the token to either A4 or B5, you are certainly going to lose the game.

From either A4 or B5, the only valid moves will force you to take the token to the 'death' cell. Now reason backward. In order to win the game you should play the game in a way such that the token is in either A4 or B5 when your rival gets his or her turn to move. You can put the token to A4 if you get your turn to move when the token is either in A3, or in B3, or in B4. Similarly, you can put the token to B5 if you get your turn to move when the token is either in B4, or in C4, or in C5. This means, your objective should be to get your turn to move when the token is in any of the following cells—A3, B3, B4, C4 or C5. Note that if the token is in column A, the only valid move is upwards. Since you need to get your turn when the token is in A3, make sure that your rival gets his or her turn when the token is in A2. Similarly, when the token is in row 5, the only valid move is leftwards. Since you must get your turn when the token is in C5, make sure that your rival gets his or her turn when the token is in D5. Figure 2.8 indicates the cells that we have identified so far as the ones from which you should move and the cells which you should force your rival to move from in order to win the game.

Now, follow a simple principle and reason backward. The principle is to make sure that you get your turn when the token

Figure 2.8

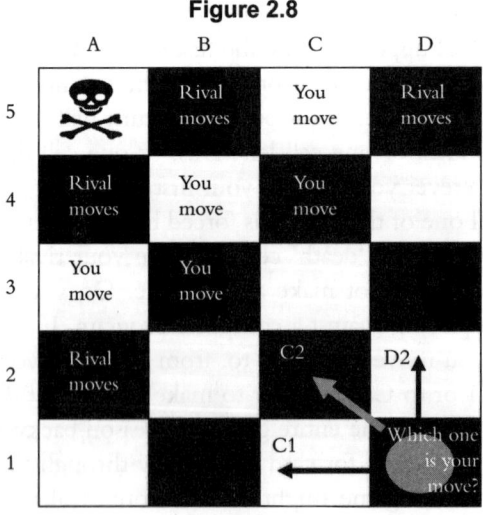

is in a cell from where any valid move will put it in a cell marked "Rival moves," and to make sure that in his or her turn the rival finds the token in cells from where any valid move will put the token in a cell marked "You move." For example, since we already identified that you will surely win if the token is in A2 when it is your rival's turn to move, you must make sure that you get your turn when the token is in A1 or B1 or B2. From either of A1, B1 and B2 a valid move can put the token in A2. On the other hand, we have already identified that you will surely win if you get your turn to move when the token is in either B3, B4 or C4. So, in the previous turn you must try to move the token to C3, which will force your rival to move the token to either B3, B4 or C4 in his or her turn. This way, reasoning backward, we can trace the game to the first move and find what should be your first move to ensure your win. Figure 2.9 indicates the cells from which you should move, and the cells which you should force your rival to move from, in order to win the game.

Now, reasoning backward through the game, we can see that in your first turn you should move the token to C1. From C1 your rival can move it either to B1, B2 or C2. If she or he puts it in B1 or B2, in your next turn you can move the token to A2 and from

Figure 2.9

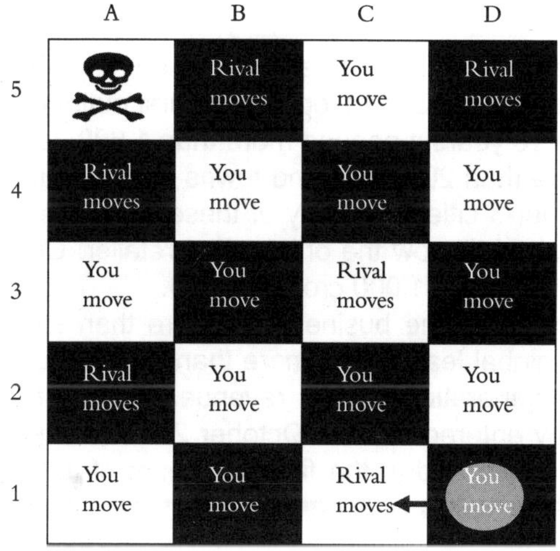

	A	B	C	D
5	☠	Rival moves	You move	Rival moves
4	Rival moves	You move	You move	You move
3	You move	You move	Rival moves	You move
2	Rival moves	You move	You move	You move
1	You move	You move	Rival moves ◄	You move

there in two more moves it will be in A4 with your rival's turn to move. If from C1 your rival put the token to C2, you should move it to C3. From C3 your rival can move it to either B3, B4 or C4, and irrespective of where she or he moves the token to you can move it to either A4 or B5 in your next turn. So, by moving the token to C1 in your first turn you ensure that you win the game, and that was achieved by looking forwards and reasoning backward. This method is technically called *backward induction*.

Representing Sequential Move Games in Extensive Form—Game Trees

An extensive form representation of a sequential move game helps in seeing through the game and reasoning backward. Let us take up a business example now, first to represent the game in extensive form and then to make decisions reasoning backwards.

This example is motivated by the case of Starbucks entering the branded café business in India, which was till then a monopoly of CCD.

Case Study 2.1: Starbucks' Entry in Indian Café Retail

Café Coffee Day (CCD), owned by Amalgamated Bean Coffee Trading Co. Ltd (ABCTCL), opened its first outlet in Bangalore in 1996. In 18 years it opened more than 1,600 outlets spread across more than 200 cities and towns across India, including tier-2 and tier-3 cities. In many of these cities the CCD outlet is the first, and till now the only coffee retailer. CCD's annual revenue exceeded ₹1,000 crore in 2014.

Starbucks is in the business for more than four decades, and is the global leader with more than 20,000 outlets across the planet generating annual revenue of more than US$15 billion. They entered India in October 2012 as a joint venture with Tata Group and in the first two years of its presence in

(Case Study contd.)

(Case Study contd.)

India opened up 58 outlets in the upscale locales of Mumbai, Delhi, Gurgaon, Pune, Bangalore and Hyderabad.

Instead of taking CCD head-on, Starbucks have been cautious with its expansion. While they opened close to 1,000 outlets in China during the same period, in India they restricted to just 58. In India they targeted the Indian counterpart of their habitual American patrons who made the company a US$60 billion darling of Wall Street. Instead of trying to poach the younger crowd that hangout at CCD outlets, in India they targeted the executives working in the business districts of Mumbai, Pune or Gurgaon who not only want a caffeine fix during lunch break or after work, but also appreciate the lounge ambiance where they can loosen their tie knots for a while and grab a bite.

Starbucks knew that they cannot beat CCD in price. CCD's business model is in control of the entire supply chain from beans to cup. ABCTCL owns more than 10,000 acres of plantation, and have a strong presence in coffee beans and ground coffee retail. As a result their average cost is low. Starbucks needed a set of customers who value the green mermaid logo and won't mind spending ₹500 for a cup of latte and a sandwich or wrap. The college going customers of CCD cannot afford that pricy cup of latte. In the tier-2 and tier-3 cities where CCD is present, there aren't enough takers of Starbucks as a lifestyle brand. So, Starbucks crafted out a market for itself without stepping on the tail of CCD.

It is CCD who retaliated by opening CCD Lounges in a bid to be present in the upscale segment. As of December 2014 they opened 43 lounges.

Source: Author.

Let us consider a generic game between two firms which we will simply refer to as the Entrant and the Incumbent. The Entrant is contemplating entry in a market where the Incumbent is a monopoly. Entrant can either take Incumbent head-on by entering with an identical product, or it may create a differentiated product and target a different segment, possibly a premium one. For ease of understanding let us name the segment where Incumbent is present as Segment-A, and the premium segment as Segment-B. In either case the Incumbent will have a response. If the Entrant enters Segment-A with an identical product, the Incumbent may retaliate by cutting price, it may create a differentiated product, or it may do nothing and be a sitting duck. Indeed there are other plausible responses like escalating advertising, expansion of outlets (in case of retail) or capacity, etc., but for the purpose of drawing the game tree we will restrict the plausible responses to three. In case the Entrant enters with a differentiated product targeting a different set of consumers in Segment-B, the Incumbent may respond by creating a similar product as that of the Entrant and expanding presence in Segment-B, or it may do nothing.

The extensive form representation of the game is given in Figure 2.10.

Figure 2.10

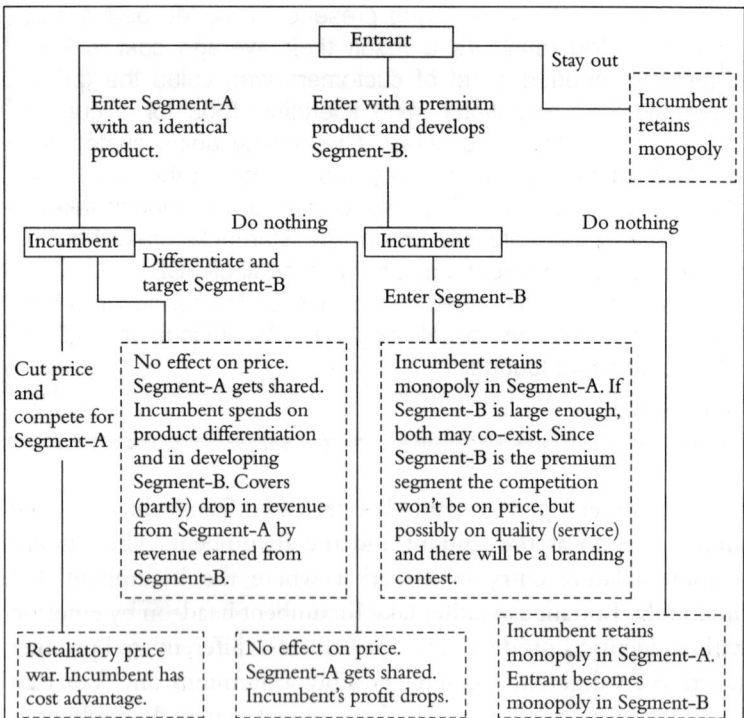

Now we can see through the game using the game tree given in Figure 2.10 and reason backward to arrive at decisions.

Putting themselves in the shoes of the Incumbent, the Entrant should anticipate what the Incumbent will do if they enter Segment-A with an identical product as that of the Incumbent. If the Incumbent chooses to do nothing, then their profits drop. So, Entrant may rule out the possibility that the Incumbent will do nothing. In the language of game theory it is called a dominated strategy, which we will discuss in length in Chapter 3. That leaves the Incumbent with two options—either cut price and compete for dominance in Segment-A, or differentiate product and target Segment-B. The later involves cost of product differentiation as well as segment development cost. The revenue earned from the premium segment may cover part of its lost revenue from Segment-A, but venturing into a new segment itself is a risky business. If the Incumbent has a strong cost advantage, cutting down price and initiating a price war will be better for them. Indeed they will lose revenue due to undercutting of price, but they are in a strong position to win the price war. Foreseeing these possibilities, the Entrant should anticipate a price war if they enter Segment-A with a product that is identical to that of the Incumbent. This was the scenario that Starbucks anticipated. They knew that CCD had a strong cost advantage owing to their business model of controlling the supply chain from bean to cup. If you don't have much chance of winning a price war, or if the cost of winning it is too high, why should you even get there? Taking CCD head-on didn't make sense to Starbucks despite their deep pocket.

If the Entrant enters with a premium product and develops Segment-B, the Incumbent might either do nothing or might want to have a share of the pie in the premium segment. The consumers in the premium segment are less sensitive to price and appreciate quality and brand value. So, if the Entrant develops Segment-B and if the Incumbent also steps in the premium segment, retaliating with price cut is not an option for the Entrant. Foreseeing that, the Incumbent should want to be present in the premium segment too. In this scenario, Incumbent's revenue from Segment-A remains unaffected. If they can at least break-even in Segment-B, why shouldn't they be present in Segment B too?

Indeed there will be a branding contest but there might be room for both in the premium segment, as is the case in the upscale café business in India. In that case the Entrant should enter Segment-B with a premium product, rather than stay out.

Putting Payoffs in the Game Tree

It always helps if we have some numbers to compare while making decisions. In a game, what the players play for is called payoff. Payoff is a generic term, and it may or may not be economic. It may very well be something abstract like 'happiness' or 'wellbeing'. In business games though, payoffs are economic variables and are measurable. However, what a firm sees as payoff depends on its strategic planning. It might be operating profit, contribution or revenue.

It is possible to forecast demand under different comparative scenarios. Hence, the firms can calculate their own as well as rival's projected revenue figures for different prices. Firms know their own costs and with a narrow margin of error can estimate the competitors cost. So they can calculate their own as well as rival's projected contribution or profit figures too. Calculating or projecting payoffs is not within the scope of game theory and hence is not within the scope of this book. The following example will help in understanding how payoffs are formed.

Amifab Co. Pvt. Ltd and BK Industries are the only suppliers of canvas fabric to backpack manufacturers in a particular region of North India. Amifab is the price leader, and every month BK Industries post their price for the month after observing the price posted by Amifab. Sales data for the first 11 months of 2014 are summarized in Table 2.1.

There was a rise in the cost of production since the beginning of July and Amifab figured that they were left with a very thin margin of ₹1.9 per sq. metre. Cost per square metre of output for Amifab is given in Table 2.2.

Since September, Amifab increased the price from ₹100 per sq. metre to ₹110. Being sure that BK's cost of production is almost same as their cost, Amifab expected BK to follow suit and

Table 2.1: Sales Quantity and Price of Canvas for 2014

Month	Amifab		BK Industries	
	Price (₹ per sq. metre)	Sales (sq. metre)	Price (₹ per sq. metre)	Sales (sq. metre)
January	100	120,000	100	100,000
February	100	120,000	100	100,000
March	100	120,000	100	100,000
April	100	140,000	110	80,000
May	100	120,000	100	100,000
June	100	120,000	100	100,000
July	100	120,000	100	100,000
August	100	120,000	100	100,000
September	110	100,000	100	120,000
October	110	100,000	100	120,000
November	110	100,000	100	120,000

Table 2.2: Cost of Canvas—Amifab Co. Pvt. Ltd (per sq. metre)

	Production volume (thousand sq. metre)									
	60	70	80	90	100	110	120	130	140	150
Direct labour	39	35	32	29	27	26	25	26	28	30
Material	14	14	14	14	14	14	14	14	14	14
Power	30	28.5	27.5	26.5	26	25.5	25	24.6	24.3	24
General overhead	24	20.7	18.3	16.1	14.5	13.3	12	11.2	10.4	9.7
Admin and selling	44	37.8	33.2	29.4	26.5	24.2	22.1	20.4	18.9	17.7
Total	151	136	125	115	108	103	98.1	96.2	95.6	95.4

raise price from ₹100 per sq. metre to ₹110. But for three months since September BK held on the price of ₹100 showing no sign of increasing it. What went wrong in the calculation of Amifab? To get an explanation let us first put the pricing game between Amifab and BK in extensive form. The game tree is given in Figure 2.11.

Figure 2.11

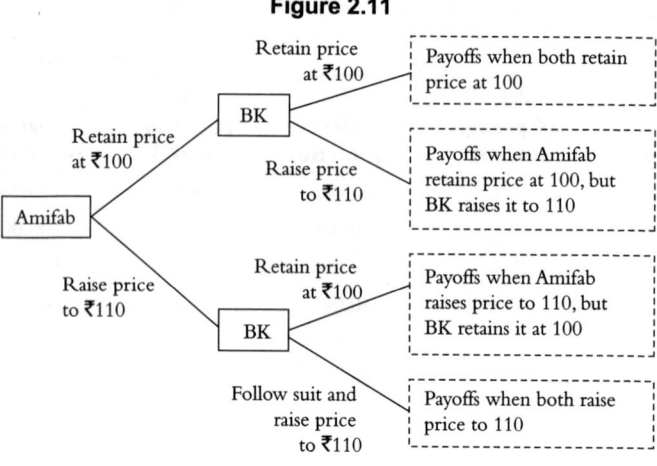

In order to anticipate BK's response to their pricing decision, Amifab needs to have an estimate of BK's payoffs. Let us suppose that payoffs are operating profits for both firms. To arrive at the profit figures under different price scenarios we need the sales figures. There are four price scenarios—(₹100, ₹100), (₹100, ₹110), (₹110, ₹100) and (₹110, ₹110), as shown in Figure 2.11. The first figure in the parenthesis indicates the price charged by Amifab, who is the first mover in the above-mentioned game, and the latter figure indicates the price charged by BK, who is the second mover. From the data given in Table 2.1 we get the sales quantities for three of the price scenarios except (₹110, ₹110), that is, when both raise price to ₹110. Note that when BK raised price to ₹110 in April while Amifab retained at ₹100, BK sold 80 thousands sq. metres of canvas in spite of its higher price. That means BK has a set of loyal customers who do not switch supplier because of ₹10 rise in price, and this set of customers provide BK Industries with a sales quantity of 80 thousand sq. metres per month. Similarly, we can see that Amifab too has a set of loyal customers who provide them with a sales quantity of 100 thousands sq. metres every month and do not switch suppliers for a difference in price of ₹10. Though Amifab charged ₹110 since September while BK retained price at ₹100, Amifab sold 100 thousands sq. metres.

During the period from January till November, aggregate monthly sales remained stable at 220 thousand sq. metres. This aggregate quantity includes the demand from the loyal customers of the two firms, which sums up to 180 thousand sq. metres. The remaining demand of 40 thousand sq. metres comes from price-sensitive customers who buy from the firm that charges ₹100 per sq. metre. When both firms charge ₹100 per sq. metre, this demand from price seekers gets equally divided. If both firm increase price to ₹110, some of the price-seeking customers will switch to some cheaper substitute of canvas, resulting in a drop in demand. The magnitude of this demand attrition can be measured using forecasting techniques. Suppose Amifab finds that aggregate demand will drop by 20 thousand sq. metres if both firms price at ₹110. This demand attrition will be entirely due to the price seekers who will substitute canvas by some cheaper material. So, in the (₹110, ₹110) price scenario the price-seeking customers will demand only 20 thousand sq. metres, which will be equally shared between the firms. The demand from the loyal customers will not get affected. Hence Amifab will be able to sell 110 thousand sq. metre of canvas, of which 10 thousand sq. metres is demanded by the price-sensitive customers, and BK will be able to sell 90 thousand sq. metres, of which 10 thousand sq. metre of demand comes from the remaining price-sensitive customers. Sales figures of each firm, under all four price scenarios, are summarized in Table 2.3.

It is safe to assume that the production costs of the two firms are comparable, and the cost data of Amifab can be used as a proxy for cost figures of BK. With that assumption Amifab should be able to calculate their operating profits as well as BK's operating profits under different price scenarios. The calculations are summarized in Table 2.3.

Now, we can put the payoffs in the game tree. The game is represented in extensive form with payoffs in Figure 2.12.

In the extensive form representation of the pricing game given in Figure 2.12, the payoffs are given as figures in parenthesis at the end of each branch of the game tree. For example, if Amifab prices at ₹100 and BK too prices its product at ₹100, Amifab earns an operating profit of ₹228,000 and BK loses ₹800,000. The payoffs

Table 2.3: Sales and Operating Profit under Different Price Scenarios

	Price per sq. metre (₹)	Sales (thousand sq. metre)	Cost per sq. metre (₹)	Operating margin (₹)	Operating profit (₹ in thousand)
Amifab	100	120	98.1	1.9	228
BK	100	100	108	−8	−800
Amifab	100	140	95.6	4.4	616
BK	110	80	125	−15	−1200
Amifab	110	100	108	2	200
BK	100	120	98.1	1.9	228
Amifab	110	110	103	7	770
BK	110	90	115	−5	−450

Figure 2.12

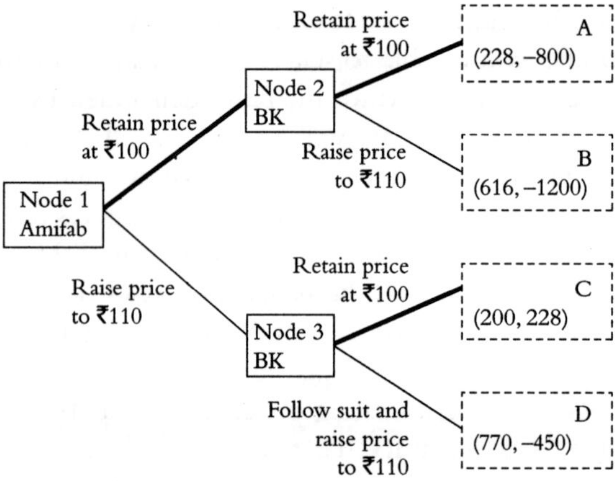

are indicated as (228, −800) in box A, which is placed at the end of the branches that indicate that both Amifab and BK chose ₹100 as their respective prices. The first figures in the parentheses are Amifab's payoff and the second one, BK's.

With the payoffs in place, it is now easy to see through the game and reason backward. Comparing BK's payoffs in boxes A and B,

Amifab should anticipate that BK will choose price ₹100 at node 2, that is, Amifab should anticipate that BK will price at ₹100 if Amifab chooses to retain price at ₹100. Similarly, comparing BK's payoffs in boxes C and D, Amifab should anticipate that BK will choose price ₹100 at node 3 too, that is, Amifab should anticipate that BK will price at ₹100 even if Amifab chooses to raise price at ₹100. So the branches leading to boxes B and D may be ignored. The branches that BK will choose at node 2 and at node 3 are the thick ones. Foreseeing that BK will choose the thick branches at node 2 and 3, Amifab should anticipate that the game will end either in box A or in box C, depending on their choice at node 1. Comparing their own payoff in box A vis-à-vis that in box C, Amifab should choose the thick branch in node 1, that is, anticipating that BK will choose to keep price at ₹100 irrespective of whether Amifab increases it to ₹110 or not, Amifab should have retained price at ₹100 only. This method of solving a sequential move game using backward reasoning is called backward induction. Had they applied this method in decision-making, Amifab would have not increased price to ₹110.

The branches along which the game unfolds when rational players play the game and make a correct decision at each node is called the equilibrium path of the game. In Figure 2.12 the equilibrium path of the pricing game consists of the thick branches that connect node 1 to box A via node 2. The payoffs given in box A are the equilibrium payoffs of the game. In the next section we will explore some games where there exists first mover's advantage.

First Mover's Advantage

In sequential move games the sequence of moves matter. Sometimes it is advantageous to move first. If a player's equilibrium payoff is more when the player moves first, vis-à-vis when the same player moves after the rival, then there exists first mover's advantage. To understand first mover's advantage, let us explore a game of new product development. The game scenario is motivated by the case of development of super-jumbo jets.

Case Study 2.2: Boeing, Airbus and the Ultra-high-capacity Airliner

In 1990, B747-400 was the largest passenger aircraft and was known as the jumbo jet. But the price of aviation fuel was increasing and the airlines were looking for a more fuel-efficient alternative. The high-capacity carrier was required for long-haul flights as it reduces the average cost per passenger for the airlines companies. By flying high-capacity carriers on busy routes, the airlines save on take-off and landing costs. Also fuel cost per passenger comes down. Airlines were interested in the ultra-high-capacity airliner (UHCA), a super-jumbo with 600 to 800 seat capacity, that will be more fuel efficient than B747. Research revealed that demand will be small. Only a select few airlines were interested only for the busiest long-haul routes.

That was the problem. Boeing and some companies belonging to Airbus consortium conducted a feasibility study in January 1993. The study revealed that if both Boeing and Airbus develops the super-jumbo jet and share the demand, both will make loses due to lack of scale. Launching a new product in the aircraft manufacturing industry involves huge fixed costs that are sunk in nature. The projected development cost for what they started calling Very Large Commercial Transport (VLCT) was US$15 billion. Unless the manufacturer produces sufficiently high numbers of aircrafts, their average fixed costs will be too high resulting in loses. However, if only one company develops the VLCT, that company will make profits.

In January 1993 itself Boeing declared in *Business Week* that they are developing a carrier of capacity 600–800 seats, which they claimed to be the "biggest and most expensive airliner ever." Airbus was pursuing its own VLCT project and in June 1994 announced its plan to develop A3XX, which will later come to be known as A380. Boeing, in the meantime, cancelled its VLCT project and moved on with plans to develop the Dreamliner.

Source: Author.

Figure 2.13

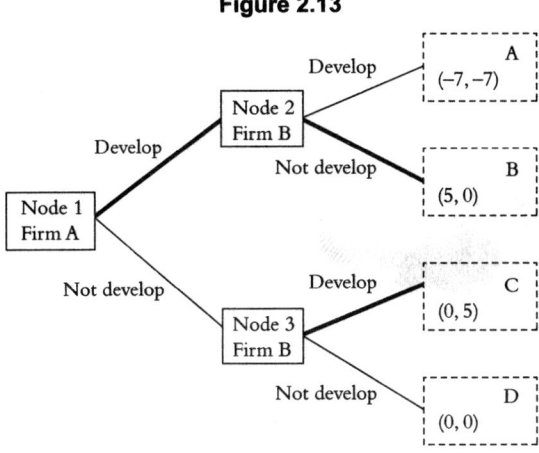

If both Airbus (Firm A) and Boeing (Firm B) develops the super jumbo, each of them will spend US$15 billion and each will make US$8 billion over a horizon of 10 years. However, if only one of them develops the super jumbo, they will spend US$15 billion but will make US$20 billion while the other stays out of this product category.

The sequential move game in extensive form, with Firm A moving first, is given in Figure 2.13.

Firm A can foresee that at node 2 Firm B will not develop, and in node 3 it will develop. Hence in node 1, Firm A's decision is to develop. By moving first and choosing to develop, if Firm A can pre-empt Firm B from developing, it nets US$5 billion while Firm B stays out of the product category. However, the same is true for Firm B if they move first and pre-empts Firm A from developing the super jumbo.

Extensive form representation of the sequential move game with Firm B moving first is given in Figure 2.14. Firm B can foresee that at node 2 Firm A will not develop, and in node 3 it will develop. Hence in node 1, Firm B's decision is to develop. Note that in Figure 2.14, the first payoffs are those of Firm B as Firm B is the first mover. By moving first Firm B nets US$5 billion and keeps Firm A out of the product category.

Figure 2.14

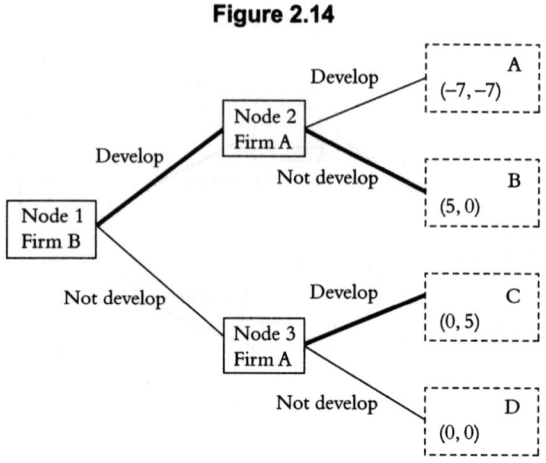

This is a symmetric game and for both players it is true that they earn more when they move first vis-à-vis when they move later. Whoever moves first nets US$5 billion and keeps the competitor out of the product category.

Indeed both firms will want to move first in such games where there exists first mover's advantage. Boeing wanted to pre-empt Airbus from forging ahead with its plan to develop a UHCA and hence declared about its plan to develop the "biggest and most expensive airliner ever." However such statements are not credible in themselves. When Airbus went ahead and presented a more credible plan about developing A3XX, Boeing backed out and dumped its plan of developing a super jumbo.

We need to address the issue of credibility of strategic moves. Chapter 5 will address the issue, and while addressing the issue we will revisit Case Study 2.2. But before that we need to discuss situations when the players make decisions without knowing the exact decision taken by the competitor. Chapters 3 and 4 discuss such games, known as simultaneous move games.

3

Prisoner's Dilemma

n a sequential move game, the players make decisions with perfect knowledge about the history of the game. However, that might not be the case in many game scenarios. When players choose actions without knowing what the rivals have done, the situation becomes identical to a game wherein the players choose actions simultaneously at the same point in time. Such games are known as simultaneous move games. In this chapter we will explore a particular class of simultaneous move games and the problems associated with simultaneity of moves in such games. When we say simultaneity of moves, we refer to the imperfection of information. In real time the decisions could be made in isolation, at different points in time, but as long as the decisions are made without knowledge of exact actions taken by rivals, the game will be classified as simultaneous move game.

Representing a Simultaneous Move Game— Payoff Matrix

Typically, a simultaneous move game is represented by a payoff matrix. Such representation is known as the strategic form representation of simultaneous move games. In order to understand the strategic form representation, let us use the example of prisoner's dilemma, a simultaneous move game formalized by A. W. Tuker (Rapoport and Chammah, 1965) which is probably the most famous game discussed under the sun.

The Story of Prisoner's Dilemma

The story of prisoner's dilemma goes something like this. It was one fine afternoon in May 1878. The Russo-Turkish war just ended and Russia's relation with Austria-Hungary was strained. A musician named Vladimir Tschesnokoff boarded a train from Belorussky terminal, Moscow. He was travelling to Innsbruck to play violin backstage where the ballet Swan Lake, composed by Tchaikovsky, was being staged. Vladimir was detained for possessing two pages of music, which were thought to be spy codes. When interrogated, Vladimir told the police that the music was composed by Tchaikovsky, who lived in Saint Petersburg. After exchanging some telegraphic communication with the law enforcers at Saint Petersburg, the police informed Vladimir that Tchaikovsky was also detained and was being interrogated at Saint Petersburg. Vladimir and Tchaikovsky didn't know each other and were effectively strangers. Incidentally, the police got hold of a person named Tchaikovsky in Saint Petersburg, but that person was not the famous composer either. Baseline situation was that two suspects, who didn't know each other, were being interrogated. There was no proof of the alleged crime, and to frame the suspects a confession from at least one of them was required. The suspects were given the following deal: They could either confess or not. If both did not confess, each would be sentenced for one year. If both confessed, each would be sentenced for five years. If one did not confess and the other confessed then the one who confessed would be set free, provided he testified against the other who did not confess. The one who did not confess would be sentenced for 10 years.

This, apparently a complex deal can be summarized in a simple matrix form, as given in Figure 3.1.

Figure 3.1

Tchaikovsky

		Confess	Not confess
Vladimir	**Confess**	Both are sentenced for 5 years. A	Vladimir is set free and Tchaikovsky is sentenced for 10 years. B
	Not confess	Vladimir is sentenced for 10 years and Tchaikovsky is set free. C	Both are sentenced for 1 year. D

In the strategic form representation given in Figure 3.1, Vladimir is the row player and Tchaikovsky is the column player. The actions available to the row player are given in the two rows, and the actions available to the column player are given in the two columns. Here, row 1 indicates that Vladimir chose the action 'confess' and row 2 indicates that he chose the action 'not confess'. Similarly, column 1 indicates that Tchaikovsky chose the action 'confess' and column 2 indicates that he chose the action 'not confess'. The row player chooses one of the rows, and the column player chooses one of the columns. The outcome of the game, when the row player chooses row 1 and column player chooses column 1, is shown in cell A. In the prisoner's dilemma game, if both Vladimir and Tchaikovsky choose the action 'confess', both are sentenced for five years, as indicated in cell A of Figure 3.1. Similarly cell B shows the outcome of the game when the row player chooses row 1 and column player chooses column 2, that is, when Vladimir chooses the action 'confess' and Tchaikovsky chooses the action 'not confess'. The outcomes of row 2–column 1 and row 2–column 2 are shown in cells C and D, respectively.

Figure 3.1 merely is a strategic form representation that shows the different outcomes contingent to different combinations of actions chosen by the two players. To convert it into a payoff matrix we need to put the payoffs in the cells. It is easy in case of prisoner's dilemma. We can simply take number of years lost in prison as payoffs. Indeed, the payoffs will either be in negative or

Figure 3.2

Tchaikovsky

	Confess	Not confess
Confess	−5, −5 A	0, −10 B
Not confess	−10, 0 C	−1, −1 D

Vladimir

zero here. The payoff matrix for prisoner's dilemma is given in Figure 3.2. In each cell there are two payoffs given. The first one is the payoff for the row player and the later one is that for the column player. For example, in cell A, when both get five years sentence, their payoffs are −5 each. In cell B, when Vladimir chooses to confess and Tchaikovsky chooses not to confess, Vladimir's payoff is 0 as he would be set free, and Tchaikovsky's payoff is −10 as he will be sentenced for 10 years. Similarly, the payoffs (−10, 0) in cell C indicate that Vladimir will be sentenced for 10 years while Tchaikovsky would be set free, and payoffs (−1, −1) in cell D indicate that both will be sentenced for one year.

It is important to note that in a simultaneous move game the players are aware of the possibilities and hence they know the payoff matrix. However, while choosing an action, they don't know what their rival chose.

Strictly Dominant Strategies

From the payoff matrix given in Figure 3.2 we can see that if Tchaikovsky chooses 'confess', Vladimir is better off choosing 'confess'. When he chooses 'confess', he gets a sentence of five years, whereas by choosing 'not confess', he gets a sentence of 10 years. In Figure 3.2, Vladimir's payoff is −5 in cell A vis-à-vis −10 in cell C. Again if Tchaikovsky chooses 'not confess', Vladimir is still better off choosing 'confess'. His payoff in cell B is 0 vis-à-vis −1 in cell D. Since Vladimir is better off choosing 'confess', irrespective of whether Tchaikovsky chooses 'confess' or 'not confess', 'confess' is his strictly dominant strategy. Similarly, Tchaikovsky is always better off choosing 'confess' irrespective of what Vladimir chooses. So, 'confess' is a strictly dominant strategy for Tchaikovsky too. In the prisoner's dilemma game, both the players have strictly dominant strategies.

A strategy is called 'strictly dominant' if and only if the player is better off playing that strategy vis-à-vis all other available strategies, irrespective of what others do (Osborne, 2004). Is driving on the left a strictly dominant strategy? You are better off driving on the left only if everyone else drives on the left. However, if everyone else drives on the right, you are better off driving on the right. So, driving on the left is not a strictly dominant strategy. But not ramming your car into a running train on a level crossing is a strictly dominant strategy. Irrespective of whether the rail crossing gate is open or closed, and irrespective of what other drivers do, it is not a wise idea to ram your vehicle into a running train.

Any player having a strictly dominant strategy will play it. Since in the prisoner's dilemma game 'confess' is the strictly dominant strategy for both players, both the suspects will choose 'confess' and each of them will receive a sentence of five years. That is the Nash equilibrium of the game. It is interesting to note that there is no dilemma in the prisoner's dilemma. The existence of strictly dominant strategies helps the suspects to overcome the dilemma.

Nash Equilibrium

A combination of strategies, known as strategy profile, constitutes Nash equilibrium if the strategies are best responses to each other (Nash, 1950). No player should have any incentive to change his or her action unilaterally from the Nash equilibrium. In the prisoner's dilemma game discussed previously, both players choosing 'confess' is a Nash equilibrium. When Tchaikovsky chooses 'confess', Vladimir does not gain by deviating from 'confess' to 'not confess'. If he deviates, his payoff reduces from −5 to −10.

Similarly, Tchaikovsky does not gain by deviating from 'confess' to 'not confess' when Vladimir chooses 'confess'.

In the Nash equilibrium of the prisoner's dilemma game, both the suspects get a sentence of five years. Had both of them not confessed, each would have gotten a sentence of one year, that is, they would have been better off had they not played their strictly dominant strategies. You may wonder why rational players would play their strictly dominant strategies! The reason is the absence of trust between the suspects who were strangers to each other. It is true that the total payoff is largest in cell D of the payoff matrix given in Figure 3.2. But each of the suspects should also realize that if he chooses 'not confess', the best response of the rival is to choose 'confess' and in that case his own payoff reduces from -1 to -10.

To understand the issue with trust in the prisoner's dilemma game, suppose the police allowed the suspects to meet and possibly make a pact (Poundstone, 1992). However, the game will still be simultaneous move as during the time of interrogation the suspects will be separated and they will have to make decision without knowing what the other chose to do. Can they stick to a pact of not confessing? Putting himself in the shoes of the other suspect, Vladimir should be able to conjecture that if he sticks to the pact and chooses 'not confess', Tchaikovsky's selfish best response would be to choose 'confess'. Thus, realizing that by sticking to the pact he will end up with a 10 years sentence, Vladimir will break the pact and choose 'confess'. The same is true for Tchaikovsky, and hence the suspects will confess breaking the pact.

In the prisoner's dilemma game, all players have strictly dominant strategies, and the Nash equilibrium is always arrived at as each player always plays his/her strictly dominant strategy. However, the players end up accepting lesser payoffs due to selfishness and lack of trust. This finding seems to be a bleak one, and indeed it is a bleak one for the suspects. But the Nash equilibrium outcome of the game is fascinating for the police. The learning for a practising manager is that it is bad to be playing a prisoner's dilemma, but good to make others play. Further, we will explore a business situation which is essentially a prisoner's dilemma game.

Grab the Deal—Discount Offer on Price as Prisoner's Dilemma

From a bargaining perspective, it is always better for a buyer to have multiple suppliers than to commit to a single supplier. By having the option of buying from different suppliers the buyer can make the suppliers bid against each other and get high discounts. The situation is strikingly similar to the prisoner's dilemma. Effectively, the buyer can make the suppliers play a prisoner's dilemma against each other. We can take a leaf out of the book of Enrique Dupuy, who was the chief financial officer (CFO) of Spanish airlines Iberia during the early years of the twenty-first century.

Case Study 3.1: Battleground Iberia—Boeing versus Airbus

In 2002–03, Iberia wanted to buy 12 planes replacing six of their 20-year-old Boeing 747-200 jumbo jets. They were looking for fuel-efficient wide body carriers and the options were Boeing's 777 and Airbus' A340. The new A340s could fly a bit farther and had more lifting power than the 777s. The new Boeing plane was lighter, held more seats and burnt less fuel. However, the Boeing plane, with a catalogue price around US$215 million, listed for some US$25 million more than the A340.

Enrique Dupuy, Iberia's CFO, invited Toby Bright, Boeing's top sales executive in Europe, and offered to buy 12 new 777s for its long-haul South American sector. Boeing had last sold Iberia planes in 1995, and since then the carrier had bought more than 100 Airbus jets. Once the underdog, Airbus has closed the gap from 1997, when Boeing built 620 planes to Airbus's 294, and in 2002 the European plane maker was expected to overtake its US rival. Having worked as Boeing's chief salesman in Europe, which is Airbus's home turf, Toby Bright had heard similar offers from customers who eventually bought Airbus planes. So he wondered if he is being brought in as a stalking horse. Yet replacing Iberia's old 747s with new 777s would be Boeing's last chance for years to win back

(Case Study contd.)

(Case Study contd.)

Iberia. Iberia was one of the industry's few highly profitable carriers, thanks to a thorough restructuring before the national carrier was privatized in early 2001, and one of the few airlines who were financially healthy enough to be able to order new planes. So, it was an opportunity for Toby Bright to get a toehold in the European market.

Enrique Dupuy was game for hard bargain and asked for discount exceeding 40 per cent. He threatened Boeing that Iberia might go for an all-Airbus fleet, which would make Iberia's operations simpler and cheaper as switching back to Boeing would require big investments in parts and pilot training. Dupuy knew that going all-Airbus might weaken Iberia's hand in future deals. He contacted Airbus' John Leahy too and asked for 40 per cent discount. He said he'll get Boeing offer 50 per cent. That was a shocker for John Leahy. Having clinched a separate deal with Iberia for three new Airbus A340 in June 2002, he thought he might bag the contract with minimal competition. But Dupuy had other plans and wanted to make John Leahy fight for the order. For Airbus, Iberia was a crucial turf to defend.

The Airbus was cheaper than the Boeing, and the A340's four engines help it operate better in some high-altitude Latin American airports. But Iberia figured that they could fit 24 more seats on the 777s and boost revenue. Also the 777 would cost 8 per cent less to maintain than the A340 and the savings would be considerable. In early November 2002, Airbus and Boeing presented initial bids on their latest planes. As negotiations began, Mr Dupuy told both companies his rule: Whoever hits its target, wins the order. The race was on.

While reporting the case, Wall Street Journal (Eastern edition), New York, dated 10 March 2003 wrote: "Airbus and Boeing may own the jetliner market, with its projected sales of more than US$1 trillion in the next 20 years, but right now they don't control it." They were rather being controlled by Enrique Dupuy. In the end, Iberia agreed to buy nine A340-600s and took options to buy three more. Airbus nosed ahead in the horse race due to lower price and its plane's common design with the rest of Iberia's fleet.

Source: *Wall Street Journal*, 2003.

The case can be seen as a prisoner's dilemma. Firm A (Airbus) and Firm B (Boeing) are the two suppliers. Indeed there was a price difference between Boeing's 777 and Airbus' A340, but for simplicity let us construct an example with two suppliers selling identical products that are identically priced. Suppose the tag price for both Firm A and Firm B is US$225 million. The buyer wants to buy 12 planes, and will buy from the supplier who offers a higher discount. In case the suppliers match discount, the buyer will split the deal. In the case of Iberia deal too, the buyer was ready to go for a mixed fleet.

We can reduce the situation into a 2 × 2 simultaneous move game by supposing that there are two possible discounts—high and low. Let high discount mean 35 per cent discount and low mean 30 per cent. In fact, in the Iberia deal, Boeing lost it by a difference of 3 per cent. The strategic form representation of the game is given in Figure 3.3.

Suppose the manufacturing cost of a plane is US$110 million. Now we can calculate the profits for each price combination and, thus, create the payoff matrix. The margin is US$36.25 million at 35 per cent discount and US$47.5 million at 30 per cent discount. When the discounts match and the deal gets split, each firm gets to sell six planes. However, by offering the higher discount, a firm can get the entire deal and sell 12 planes. The payoff matrix for the game is given in Figure 3.4. Payoffs are profits.

Figure 3.3

Firm B

		High discount	Low discount
Firm A	**High discount**	The deal gets split between the two. Each sells six planes. The profit margin is low. Profit is low for both.	Firm A gets the entire deal and sells 12 planes. Profit margin is low, but profit is high due to high sales. Firm B gets nothing.
	Low discount	Firm A gets nothing. Firm B gets the entire deal and sells 12 planes. Profit margin is low, but profit is high due to high sales.	The deal gets split between the two. Each sells six planes. The profit margin is high. Profit is moderate for both.

Figure 3.4

		Firm B	
		High discount	*Low discount*
Firm A	*High discount*	**217.5, 217.5**	435, 0
	Low discount	0, 435	285, 285

From Figure 3.4, we can see that offering high discount is the strictly dominant strategy for both firms. Irrespective of what Firm B does, Firm A is strictly better off offering high discount. The same is true for Firm B. So, in the Nash equilibrium of the game, both firms will offer high discount and get a payoff of US$217.5 million. However, if both offered low discount, they would have gained as their payoff would have increased to US$285 million. But they cannot reach an agreement without a binding clause in a contract, due to absence of trust. If Firm A offers low discount, Firm B will offer high discount and grab the entire deal. Firm A will do the same if Firm B offers low discount.

Assurance Game— a Game with a Real Dilemma

In the prisoner's dilemma scenario discussed in the section 'Representing a Simultaneous Move Game—Payoff Matrix', the suspects were given a raw deal that forced them to confess. Suppose the suspects were given a slightly softer deal. They still just have two options—either confess or not. But if both did not confess, they would be given the benefit of doubt and set free. The rest of the deal remains unchanged. If both confessed, each would be sentenced for five years. If one did not confess and the other confessed, then the one who confessed would be set free, provided he testified against the other who did not confess, and the one who did not confess would be sentenced for 10 years. As

Figure 3.5

Tchaikovsky

	Confess	Not confess
Confess	−5, −5	0, −10
Not confess	−10, 0	0, 0

Vladimir

before, the suspects cannot communicate with each other, and while choosing an action they don't know what the other chose. So it is still a simultaneous move game with two players, each having two actions to choose from. The payoff matrix of the 2 × 2 simultaneous move game is given in Figure 3.5.

With this modification 'confess' is no longer a strictly dominant strategy. If Tchaikovsky chooses 'confess', Vladimir is still better off choosing 'confess'; when he chooses 'confess' he gets a sentence of five years, whereas by choosing 'not confess', he gets a sentence of 10 years. However, if Tchaikovsky chooses 'not confess', Vladimir's payoff is 0 from choosing 'confess' as well as from choosing 'not confess'. Refer to Figure 3.5.

With the modification made in the deal, 'confess' has become a weakly dominant strategy for both the players. A strategy is called "weakly dominant" if the player is either better off or as well off playing that strategy vis-à-vis all other available strategies, irrespective of what others do (Osborne, 2004). Note that with the modification in the deal, we have lost the uniqueness of the Nash equilibrium. Now there exist two Nash equilibria. The strategy profile (confess, confess) constitutes Nash equilibrium as in the original prisoner's dilemma. But (not confess, not confess) also constitutes another Nash equilibrium. When Tchaikovsky chooses 'not confess', Vladimir does not gain by deviating from 'not confess' to 'confess'. His payoff remains at 0. Similarly, Tchaikovsky does not gain by deviating from 'not confess' to 'confess' when Vladimir chooses 'not confess'. The existence of two Nash equilibria creates a real dilemma.

Among the two Nash equilibria, (not confess, not confess) is the better one for the suspects. But, which Nash equilibrium is more likely to occur? Since 'confess' is the weakly dominant strategy, by definition that means the suspects do not gain by choosing 'not confess' instead of 'confess'. If a suspect chooses 'not confess', he might be free or he might land up in prison for 10 years depending on the choice of the other suspect. On the other hand, if he chooses 'confess', he might be free or might get a sentence of five years. So, even if there is a small chance that the other suspect might choose 'confess', it is better to choose 'confess'.

Suppose the suspects are allowed to meet and possibly make a pact. The game will still be simultaneous move as during the time of interrogation the suspects will be separated and they will have to make decision without knowing what the other chose to do. With the modified deal, it is possible for the suspects to make a pact of not confessing and to stick to the pact. Both the suspects realize that (not confess, not confess) is the win–win outcome for them. Putting himself in the shoes of the other suspect, Vladimir should be able to conjecture that if he sticks to the pact and chooses 'not confess', Tchaikovsky does not gain anything by deviating and breaking the pact. Tchaikovsky's payoff is 0 when Vladimir chooses 'not confess', irrespective of whether he chooses 'not confess' or 'confess'. Tchaikovsky should be able to make similar conjecture about Vladimir's decision. The suspects wanted to reach the (not confess, not confess) equilibrium, and required mutual assurance of choosing 'not confess' to be able to reach that win–win outcome. This modified prisoner's dilemma is known as the "assurance game." The characteristic of the assurance game is that each player has a weakly dominant strategy, and there exists two Nash equilibria. One of the equilibria offers higher payoffs to both the players and is known as the focal equilibrium (McCain, 2007) of the game. However, to reach the focal equilibrium, the players need mutual assurance that they will not choose their weakly dominant strategies. The assurance helps the players to overcome the dilemma.

How the Tobacco Industry Could Cease Advertisements on TV

After a gap of four decades, cigarette advertisement is back on television in the UK. Britain banned television ads promoting cigarettes in the 1960s, and ads for other tobacco products have been prohibited since the early 1990s. However, current advertising codes weren't designed with electronic cigarettes in mind, which is now a US$3 billion industry worldwide. The rules around e-cigarettes are still somewhat hazy and this regulatory gap has not gone unnoticed by tobacco companies, who have already spent a staggering £60 million during the four-year period since 2009, according to the market research company Nielsen. These products, including e-cigarettes and nicotine patches, are now classified as 'smoking deterrent products' in the UK. Manufacturers are now allowed to advertise the use of electronic cigarettes on TV, as long as they do not promote tobacco or target non-smokers or young people. The European Union passed new rules that starting in 2016 it will re-classify e-cigarettes as "tobacco-related products." They will be subject to the same advertising ban as regular cigarettes. Till then it is time for heavy advertising on TV. In 2014 itself, the industry spent more than £15 million on advertising electronic cigarettes. In this context, it won't be irrelevant to look back at the case of cigarette advertising on television during the 1950s and 1960s, when the tobacco companies in Europe and North America used to advertise heavily on TV.

Case Study 3.2: Tobacco Advertising on TV—USA (1950–70)

American television was once rife with cigarette advertising. A surge of advertising in the 1950s saw tobacco companies sponsoring TV shows including prime-time family cartoons like "The Flintstones," which had a captive young viewership. TV commercials for cigarettes featured stars like Lee Marvin, John

(Case Study contd.)

(Case Study contd.)

Wayne or Irene Ryan. Reports on harmful effects on smoking, particularly tobacco's connection to lung cancer and heart diseases, were already in circulation during the late 1950s and early 1960s. Nevertheless, the tobacco companies like Philip Morris, R. J. Reynolds Tobacco Co., Brown & Williamson, etc., continued to advertise cigarette brands on TV. The tobacco companies were unperturbed till the US Surgeon General Dr Luther L. Terry published a report of the advisory committee on the health hazards of smoking in 1964. The report held cigarette smoking responsible for 70 per cent increase in the mortality rate of smokers over non-smokers. The report also estimated that smokers have 9- to 10-fold risk of developing lung cancer vis-à-vis non-smokers, and identified smoking as the most important reason behind chronic bronchitis. At this point in time, the tobacco companies started fearing lawsuits. In 1964 itself, 17 tobacco liability suits were filed in the USA. Awareness on dangers of smoking gained momentum in the next few years, and in 1967 the Federal Communications Commission mandated that anti-smoking public service announcements be aired at no cost to the advertiser.

A complete ban on cigarette advertising was suggested by the Federal Trade Commission (FTC) in 1968. Broadcasters vehemently protested against the proposal as about 10 per cent of their total TV advertising revenue came from cigarette and other tobacco advertising. Tobacco companies, on the contrary, were more willing to go along with the idea of a complete ban on cigarette advertising. They believed that a voluntary withdrawal of advertisements from TV and radio would serve two purposes. First, it would provide them some immunity against federal lawsuits that might ask them to pay for the healthcare cost of tobacco victims. Second, they recognized that since all tobacco companies will be subject to the ban, they might actually save money without losing market share to the competitors. They learnt this from their experience in the UK where the ban came in 1965. But since 1962 the tobacco companies have voluntarily withdrawn cigarette

(Case Study contd.)

(Case Study contd.)

advertisements on television for all slots before 9 PM, which included prime-time slots.

Public Health Cigarette Smoking Act, which mandated a strong health warning label on cigarette packets and banned advertising on American radio and television, was introduced to Congress in 1969. President Richard Nixon signed it into a law on 1 April 1970, but it was not until 2 January 1971 that cigarette advertising ban on television came into force. The ban served the tobacco industry well, as the industry recognized. Advertising revenue for broadcasters came down by US$63 million in the first quarter of 1971, in comparison to the same quarter of 1970. During the first quarter of 1971, tobacco industry's quarterly profit increased by US$93 million.

Sources: Buchdahl, 2013; Mahdawi, 2014; http://archive.tobacco.org/ resources/history/ (last accessed January 2015).

Keeping the history of tobacco advertising on TV (USA, 1950–70) in mind, let us first construct the advertising game, which is another example of prisoner's dilemma. During the 1960s, Philip Morris and R. J. Reynolds were the two largest tobacco companies in the USA. Let us conceive a game between Firm P (Philip Morris) and Firm R (R. J. Reynolds) wherein each firm decides at the beginning of each quarter whether to advertise on television or not. They are aware of the fact that each of them has the options either to advertise or not advertise cigarettes on TV, but while making the decision for the quarter they do not know what the other firm is doing. So, the stylized situation can be seen as a 2 × 2 simultaneous move game. In order to construct the payoff matrix of the game let us use the following symmetric payoff landscape. If they don't spend on advertising, each firm gets a contribution of US$60 million in a quarter. Advertising on TV costs US$30 million in a quarter. Advertising has two effects. On one hand, it captures US$40 million from the rival, if the rival firm does not advertise. On the other hand, it creates new smokers who contribute US$10 million to the industry in a quarter. This US$10

Figure 3.6

Firm R

	Advertise	Not advertise
Advertise	35, 35	80, 20
Not advertise	20, 80	60, 60

Firm P (row label, left of table)

million goes to the firm that advertises their product, when only one of the firms advertises. However, when both advertise, the advertisements neutralize each other and the additional contribution of US$10 million from the new smokers gets shared between the firms. So when both the firms advertise, they are left with a contribution of US$35 million each, after paying for the advertisements. The payoff matrix for the game is given in Figure 3.6.

From Figure 3.6, we can see that advertising is the strictly dominant strategy for both firms. The strategy profile (advertise, advertise) constitutes the Nash equilibrium of the game and at the Nash equilibrium, each firm gets a contribution of US$35 million in a quarter. Instead, if both the firms didn't advertise, they would have gotten contribution of US$60 million each by saving the expenditure on advertising. However, the firms cannot strike an agreement of not advertising. If, for example, Firm P does not advertise, best response of Firm R is to advertise and increase its contribution to US$80 million. Foreseeing that consequence, Firm P will rather advertise and the same argument is true for Firm R too.

With the threat of advertising ban on television in the USA during the late years of the 1960s, the tobacco industry saw it as an opportunity to be able to move from the (advertise, advertise) Nash equilibrium to the superior (not advertise, not advertise) outcome. In fact, the tobacco liability suits changed the payoffs too and the game became an assurance game. Let us suppose that the industry is faced with an expected liability of US$20 million, per quarter, as a consequence of the tobacco liability suits. However, the firm not advertising gets immunity. There won't be any effect

Figure 3.7

Firm R

	Advertise	Not advertise
Advertise	**25, 25**	60, 20
Not advertise	20, 60	**60, 60**

Firm P

on the payoff of the firm not advertising. But the advertising firm's payoff reduces by US$20 million. In case both the firms advertise, the incidence of tobacco liability is equally likely for both firms, and in that case the expected liability is US$10 million to each firm. The payoff matrix for the game with these modified payoffs is given in Figure 3.7.

From the payoff matrix given in Figure 3.7, we can see that there are two Nash equilibria for the game—(advertise, advertise) as well as (not advertise, not advertise)—with the latter being the focal equilibrium of the game. For the industry, (not advertise, not advertise) is indeed the preferred equilibrium. To reach this equilibrium, the firms need to assure each other that they won't advertise. The ban proposed in 1968 made it possible for the firms to assure each other that they won't advertise cigarette on television and radio. Hence, the tobacco companies welcomed the complete ban of cigarette advertising on television. They knew that the ban will do them good.

Prisoner's dilemma and assurance games are simultaneous move games where players have dominant strategies. But there are other types of simultaneous move games where players don't have any dominant strategy. In Chapter 4, we will discuss such games.

4

Coordination and Anti-coordination Games

There are games where there does not exist any form of dominant strategy for any player. In such games there exist multiple Nash equilibria and it becomes difficult to choose one of the equilibria. The players need to coordinate on their strategies. But in a simultaneous move game the players may not be able to coordinate in absence of any communication (Cooper, 1998).

In coordination games, it is possible that one Nash equilibrium is preferred by one of the players while the other player prefers another Nash equilibrium. However, both prefer a Nash equilibrium over coordination failure and hence they want to coordinate. In anti-coordination games, the players don't want to coordinate. In such games too there exist multiple Nash equilibria and different Nash equilibria are favourable to different players. Moreover, in such games the players prefer coordination failure over the Nash equilibrium that is favourable to the rival. Hence they don't want to coordinate.

In this chapter, we will explore examples of both coordination games as well as anti-coordination games.

Coordination Games

There are two variants of coordination games—the battle of sexes (BoS) and the pure coordination game.

Battle of Sexes

Consider a situation where a man and a woman, who are in a relationship, are faced with a decision of how to spend their Saturday evening. The choice is between a very important football match, say a final of the UEFA Champions' League, and a live performance of say Vienna Philharmonic Orchestra, which is a rare occasion for music lovers. Contrary to the usual cultural stereotype, suppose that the man prefers the concert over the football match and the woman prefers the football match over the concert. However, to both of them it is important that they go out together than that each independently enjoys their preferred entertainment. To put payoffs in the game, suppose the valuation of utility to the man from the concert is ₹10,000, and that from the football final is ₹5,000. The valuations of utility for the woman from the respective entertainments are other way round. However, to each of them the cost of being alone, in monetary terms, is ₹6,000. This cost may be attributed to either the fact that they will be missing each other, or to the subsequent fight that they will have to endure for being 'selfish', or a combination of both forms of agony. To justify how the game is a simultaneous move game, suppose the couple picked up the fight in the morning and went their ways. During the day they are not picking up each other's calls as they are mad with each other. So, in the evening they have to decide whether to hit the concert hall or the sports bar where they usually go to watch matches. The payoff matrix for the game is given in Figure 4.1. The payoffs are in thousands of rupees.

From Figure 4.1, we can see that if the woman chooses 'football', the man's best response is to choose 'football', and if the

Figure 4.1

Man

		Football	Concert
	Football	**10, 5**	4, 4
Woman			
	Concert	−1, −1	**5, 10**

woman chooses 'concert', the man's best response is to choose 'concert'. Likewise, if the man chooses 'football', the woman's best response is to choose 'football', and if the man chooses 'concert', the woman's best response is to choose 'concert'. Clearly, none of the players have any dominant strategy. There are two Nash equilibria of the game. The strategy profile (football, football) constitutes best responses to each other. So does the strategy profile (concert, concert). How do we choose an equilibrium here? There are two different ways to resolve the problem of equilibrium selection. We will revisit this game and address the problem of equilibrium selection in Chapter 5 where we will discuss strategic moves. Another way of resolving the issue is through use of mixed strategy, which we will discuss in Chapter 7. For now let us just appreciate the fact that there might be game situations that generate multiple Nash equilibrium, and hence the fact that a particular strategy profile constitutes Nash equilibrium does not mean much from a practical perspective of playing the game (Gibbons, 1992).

Clearly, there exists a possibility of not reaching any of the Nash equilibria. Note that for both the players the payoff is higher in either of the Nash equilibria than on non-Nash outcomes. So, failing to reach a Nash equilibrium is sad for the players of this game. But that is very likely. Suppose the woman chooses 'football' expecting to hit the (football, football) equilibrium, and the man chooses 'concert' expecting the (concert, concert) equilibrium, they end up at the (football, concert) outcome which gives each of them a lesser payoff than what they could have obtained on any of the Nash equilibria. If the players try to be too nice to each other the outcome will be even worse. Suppose the woman chooses 'concert' expecting the man to choose 'concert', and the man chooses 'football' expecting the woman to choose 'football', they end up at the (concert, football) outcome which gives them the worst outcome in a scenario similar to the situation of Della and Jim in O. Henry's *The Gift of the Magi*. Of course, as absolutely materialistic and heartless beings, we are not counting the abstract payoff from happiness here. Game theory can work with abstract payoffs too, but if we incorporate such abstract payoffs in the BoS game, the character of the game will change. The purpose of the BoS game is to show you that in games with multiple equilibria coordination failure is possible and such a coordination failure might be costly for the players.

Game of Pure Coordination

One characteristic of the BoS is that the two players have con-flicting preferences over the two plausible Nash equilibria. The man prefers the (concert, concert) equilibrium and the woman prefers the (football, football) equilibrium. In a game of pure coordination, both players are indifferent between the two Nash equilibria (Dixit and Skeath, 2004). In order to bring out that characteristic of the coordination game, let us make a small change in the BoS game. Suppose the payoff from watching the football match together, as well as that from attending the con-cert together are ₹10,000 for both the man and the woman. Let the cost of separation be ₹6,000 as before. In order to rationalize the simultaneous move suppose they went out without finally deciding and then they are unable to communicate for some odd reason like devise malfunction or network congestion. With this modification, the payoff matrix for the game looks like the one given in Figure 4.2.

In this game too there are two Nash equilibria—(football, football) and (concert, concert), and here too coordination failure is likely and costly to the players. Here, the problem of

Figure 4.2

		Man	
		Football	*Concert*
Woman	*Football*	**10, 10**	4, 4
	Concert	4, 4	**10, 10**

coordination failure can be addressed simply through communication since both the players are indifferent between the two Nash equilibria.

Technology Adoption in Presence of Network Effect—A Coordination Game

Many modern technologies exhibit network effect and are subject to increasing returns. Such technologies remain unviable unless the number of users attains a critical mass. The number of users of a technology and its compatible technologies constitutes the network size of the technology. Network effect refers to the phenomena that the user's utility increases with increase in network size, or user's cost decreases with increase in network size (Shy, 2004).

The choice of the operating system and software applications by users of personal computers highlights the problem of technology adoption in presence of network effect (Katz and Shapiro, 1994; Shy, 1996). Broadly, a computing system consists of the hardware, operating system and software applications. The network effect in personal computing systems arises from the fact that people use systems not only for processing and storing information, but also for sharing information. Information processed in a particular application will not open in another application unless the two applications are compatible. Now many applications use cross-platform file format and open on Windows PC, as well as on Macintosh machines. But that was not the case 20 years ago. Microsoft's growth story in the 1990s critically hinged on exploitation of network effects, and on of strategic incompatibility of MS Office with Mac OS. Mac users didn't have an option to use MS Office. For word processing, Mac users typically used Nisus Writer, a word processor supported by Mac platforms. If a Mac user composed a text in Nisus Writer and shared it with a Windows user, she could open the document in a Windows PC and read the text but the style information would be lost. Being aware of this issue and knowing that most personal computer users used MS Office, while making purchase decisions new users chose PCs with Windows interface over Macintosh and that helped Windows become the largest selling operating system for personal computers.

Now consider the following hypothetical situation faced by two business partners in 1990s. A and B were two young business school graduates who started a consulting firm in 1994. Since they needed to work together they needed to share documents and spread sheets. A used a PC with Windows and was comfortable in using MS Office. B was a Mac user and used Nisus Writer, Mariner Calc and Mariner Write, which were then incompatible to Windows platform. So, one of them needed to switch. Switching from one technology to another involves cost. Whoever switched would have to invest in a new system and also get used to the new system. Let the switching cost be ₹50,000 in 1994 and the life of a PC was four years. Let the expected net present values of profit from the business calculated for a horizon of four years was ₹800,000, which was to be shared between the two partners. But unless they used same systems they could not do business efficiently and could not earn that sum. If they didn't adopt the same computing system they would have lost 50 per cent of their businesses due to inefficiency. That basically translated into reduction in the expected net present values of profit from the business from ₹800,000 to ₹400,000 for the horizon of four years. For now let us just see the situation as a simultaneous move game. We will return to this game and discuss strategic moves in the next chapter. With the payoffs in place we can now

Figure 4.3

B

	Windows (switch)	Mac (do not switch)
A Windows (do not switch)	**400, 350**	200, 200
Mac (switch)	150, 150	**350, 400**

represent the situation in strategic form. The payoff matrix for the game is given in Figure 4.3. Payoffs are in thousands of rupees.

The game is a BoS. The two Nash equilibria are (Windows, Windows) and (Mac, Mac). The former is reached if B switches incurring a cost of ₹50,000 and the latter is reached when A switches incurring the same cost. Hence, A prefers the (Windows, Windows) equilibrium and B prefers the (Mac, Mac) equilibrium. For both players the payoffs are lower when they fail to coordinate. In a simultaneous move game, the coordination failure is possible.

With this understanding of the BoS game let us now explore how coordination failure might impede adoption of a superior new technology—a phenomenon known as "excess inertia."

RFID Technology in Retail Supply Chain— Issue with Case Tagging

Radio Frequency Identification (RFID) technology is a technique of object identification and data capture, using electronic labelling and radio waves. Widespread application of the technique began during the first decade of twenty-first century in domains ranging from lending systems in libraries and monitoring of scientific experiments to baggage handling at airports and product tagging in various industries. For product tagging it is considered to be the next stage in barcode evolution.

Implementation of RFID for product tagging requires the container to bear an RFID tag, which consists of a chip that contains product information and an antenna that transmits the information. The information transmitted by the RFID tag is captured by a scanner or reader, which is connected to a computer system. Whenever a tag passes by a scanner the data is captured and transferred to the computer systems. In applications where it generates huge volume of data, a middleware known as savant is used to filter the data before transferring it to backend IT system.

There is a huge scope of increasing efficiency through implementation of RFID in retail supply chain. Typically, the retail supply chain involves the manufacturer, the retailer's distribution centres (DC), the backroom of the retail outlet and the selling floor. Product information (quantity, variety, price, date of manufacturing, date of expiry, etc.) needs to be collected and stored for future use in possible Enterprise Resource Planning (ERP) implementation when the containers enter the manufacturer's warehouse, when it leaves the warehouse and shipped/trucked to the retailer's DC, when it enters the DC, when it is trucked out of the DC for retail outlets (supermarkets, hypermarkets, etc.), when it enters the backroom of the retail outlet and when it leaves the backroom and enters the shopping floor. Using ultra-high frequency radio waves, which have a read range of more than four metres (less in Europe due to power transmission restrictions), RFID makes this data collection easier and cheaper at different nodes of the retail supply chain. The major benefits of RFID implementation for the supplier include increased labour productivity in loading and storing and from automated scanning, increase in inventory accuracy through free flow of information, automatic reporting of shipment data, improved goods transfer and payment process. For the retailer, the benefits accrue from improved labour productivity at the DC, improved efficiency in receiving and paying at the DC, reduced inventory at the DC as well as at the store backroom, reduced truck idle time, improved replenishment resulting in increased sales, decreased obsolescence or expiry and decreased theft.

Metro AG of Germany, the fifth largest retailer in the world by revenue, was one of the first to implement RFID in retail supply chain back in 2004. They encountered a BoS game when they wanted to switch from pallet level tagging to case-level tagging. Refer to Case Study 4.1.

Case Study 4.1: RFID Tagging at Metro AG

Metro group rolled out the RFID implementation to the goods received in a few selected test stores and DC in 2004. The results were positive and created a necessity for group wide rollout, as the magnitudes of relative benefits (percentage changes) were projected to be much larger due to increasing returns from the technology. RFID was implemented in Galeria Kaufhof (department stores), Real stores (hypermarket) and in Metro Cash & Carry (wholesale) using pallet tagging. In the beginning only a few large suppliers including Gillette, Nestle, Henkel, Esprit, etc., participated.

Metro started the rollout with pallet-level RFID tagging. As the pallets were unloaded from trucks and entered the DC they could be registered and checked for delivery completeness within seconds through automated barcode scan. Adoption of RFID enabled automated barcode scan at the DC even during order assembly for store shipment. This automated barcode scan resulted in increased labour productivity for Metro DCs. As per case studies done by Metro AG, total benefit from RFID implementation was estimated at €0.16 per pallet.

Similar benefits from increased labour productivity accrued to the manufacturer from automated scan during loading on trucks at the manufacturer's warehouse. Since the pallets were scanned during loading, RFID implementation also eliminated the need for manual supervision during loading both at the manufacturer's end as well as at the DCs. For the manufacturers there was further increase in labour productivity resulting from elimination of a process that involved assembling pallets in a special area before sending them to the loading area. Metro AG case studies estimated that a manufacturer

(Case Study contd.)

(Case Study contd.)

---

sending 15 trucks per day (each carrying 36 pallets) will save €40,000 per year. They assumed 250 working days in a year and wage rate of €25 per hour. This translates into a savings of €0.5 per pallet for the manufacturer. The Gen-1 RFID tags cost approximately €0.25 in 2005. Data could be written only once in these tags, but could be scanned multiple times. So, the manufacturer could reuse the tags for dispatching identical pallets. The manufacturers also needed to invest in printers. RFID printers came for around €3,000.

Metro soon identified that the benefits from RFID implementation will be a lot more for them if they could ensure a case-level tagging. Firstly, it would eliminate the need for counting the cases upon receiving the delivery at the DC. At the DC, upon receipt of an electronic order from a store, picking orders are generated and electronically forwarded to employees called pickers. Before RFID implementation, the pickers entered and confirmed the number of picked cases. Case-level RFID implementation was estimated to save four minutes per pallet for the pickers. Elimination of need for inventory count would also increase labour productivity. Metro figured out that the case-level tagging will not only increase labour productivity, but will also result in better inventory management both at the stores as well as in the DCs. Better on-shelf status was expected to result in increased sales and hence increased profit for both Metro stores as well as for the manufacturers. Metro estimated a gain of €0.09 per case for Metro group, which translates to approximately €5.5 per pallet.

Implementation of case-level tagging required training of employees at the Metro stores and DCs, as well as in the manufacturer's factories and warehouses. Metro also wanted to switch to Gen-2 tags. The Gen-2 tags were cheaper and rewritable, enabling more flexible reuse. But using the Gen-2 tags required reinvestment in the scanners. Case-level tagging also meant that the manufacturer would need a lot more tags. On an average each pallet contained 60–80 cases. A switch from pallet tagging to case tagging would require the manufacturer to use 60–80 times more tags, which was a

---

(Case Study contd.)

(Case Study contd.)

considerable cost. On the other hand, most of the benefits from the switch accrued to Metro stores and DCs. Case study done by Metro AG showed that case-level scanning will result in increased profit of €1,280,000 per year for a representative manufacturer—a benefit of €0.07 per case. The figure was not very convincing to the manufacturers. Also they were sceptical about the future of the RFID technology. If the tags became obsolete soon, they would again have to incur costs. The manufacturers, except for a few big ones, were reluctant to switch to case-level tagging.

Source: RFID: Uncovering the Value; Metro AG Future Store Initiative, 2004

It was not only Metro who encountered a situation where the suppliers were reluctant to switch to a new technology. Walmart faced the same issue while trying to implement RFID. The issue is generic for many contexts of technology adoption. Here, we are exposing the situation with a strategic form representation of the case-level RFID implementation game between a retailer and a supplier. In this chapter, we will not address the strategic moves required to deal with the situation. For that purpose we will revert back to the issue in Chapter 5.

The game is between a retailer and a supplier. Payoffs are benefits per pallet net of cost, given in Euro. These payoffs are not exact figures, but intuitively derived from the studies done by Metro. There are four possible outcomes of the simultaneous move game—both the supplier and retailer continue with pallet tagging, supplier continues with pallet tagging but retailer implement system for case tagging, both move to case tagging and the unlikely scenario where supplier adopts case tagging and retailer continues with the pallet-tagging system. The benefits per pallet (in Euros) to the supplier and retailer under each of the scenarios are summarized in Table 4.1.

Under pallet-tagging system, tag costs per pallet to the supplier was €0.025. The average fixed cost to the supplier was also €0.025

Table 4.1: Benefits (per pallet) of RFID Implementation

Outcome	Supplier's Benefit (€)	Retailer's Benefit (€)
Both continue with pallet tagging	0.50	0.16
Supplier continues with pallet tagging but retailer implement system for case tagging	0.20	0
Both move to case tagging	1.80	5.40
Supplier adopts case tagging and retailer continues with pallet tagging system	0.2	0

per pallet. The retailer only had to bear a fixed cost. The average fixed cost to the retailer was only €0.06 per pallet. The tag cost per pallet borne by the supplier was projected to increase to €1.32 if they adopt case tagging. Apart from that the supplier was required to re-train their employees resulting in an expected cost of €0.05 per pallet. Average fixed cost to the supplier was projected at €0.1 per pallet when they adopt case tagging. The retailer was also required to retrain their employees and their training cost was estimated to be €0.2 per pallet, and their fixed cost of implementing the new case tagging system was projected at €0.16 per pallet.

The payoff matrix of the simultaneous move game is given in Figure 4.4. Though the game is not symmetric, it is still a BoS game. There are two Nash equilibria. In one of the equilibria, the supplier and the retailer coordinate on pallet-tagging system and in the other they coordinate on case-tagging system with the

Figure 4.4

Retailer

Supplier		Continue with pallet-level RFID system	Switch to case-level RFID system
	Continue pallet tagging	**0.45, 0.10**	0.15, −0.36
	Switch to case tagging	−1.27, −0.06	**0.33, 5.04**

supplier preferring the former and the retailer preferring the later. A coordination failure results in loss of payoff for both and hence avoidable. We will address the issue of ensuring coordination in the next chapter.

Anti-coordination Games

So far, in this chapter and in Chapter 3, we came across two classes of games—games with players having dominant strategies and those where players don't have any dominant strategy. We also understood the role of coordination among players in the later class of games. In this section, we will discuss another kind of games belonging to the class of games where the players don't have any dominant strategy. In this kind of games the interests of the players are more starkly conflicting and they don't want to coordinate at all.

Hawk–Dove Game

The name hawk–dove was coined by evolutionary biologists J. M. Smith and George Price (Smith and Price, 1973) in the context of a behavioural game between two animals who are in conflict over a scarce resource, which may be either food or a mate. They can behave like a hawk, that is, fight for sole ownership of the resource, or behave like a dove, that is share the contested resource. If one chooses to be hawk while the other chooses to be a dove, the one who chose to be a hawk wins the resource. If both choose to be doves they share the resource. If both choose to be hawks they fight a bloody war, which is the worst outcome for both. The payoff matrix of the game is as given in Figure 4.5. Let us name the two animals as A and B.

In the hawk–dove game too, like the BoS game, there is no dominant strategy for any of the animals. If A chooses to be a hawk the best response of B is to be a dove, and vice versa. On the other hand if A chooses to be a dove the best response of B is to

Figure 4.5

	B	
	Hawk	*Dove*
Hawk	–5, –5	10, 0
Dove	0, 10	5, 5

(A labels the rows on the left side)

be a hawk. There are two Nash equilibria here—(hawk, dove) and (dove, hawk). Indeed A wants the former and B the latter. Both the Nash equilibria have the win–lose character, which highlights the innate conflict in this game.

Chicken Game

Chicken game is another variant of anti-coordination games that is more commonly used in the literature of economics, politics and business. Two drivers take up this challenge of driving towards each other on a collision course. Whoever swerves chickens out and loses the game. If both swerve, none wins. But if both dare not to swerve they collide, which is the worse outcome in this game. The payoff matrix is given in Figure 4.6. The two Nash equilibria are (dare, chicken) and (chicken, dare).

Figure 4.6

	Driver 2	
	Dare	*Chicken*
Dare	–5, –5	1, 0
Chicken	0, 1	0, 0

(Driver 1 labels the rows on the left side)

In both hawk–dove and chicken game, the players need to convince the rival that they are committed to the winning strategy, that is dare in chicken game and hawk in hawk–dove game. They can decide to close down the option to choose the other strategy, which is a standard military practice. Alternatively, they can signal to the rival that they are committed to the winning strategy. In situations where the players play similar games with different players a large number of times they may develop a reputation of daring to choose the winning strategy. We will revisit this class of games and address the issue of conflict resolution in Chapters 5 and 7.

Boeing, Airbus and a Chicken Game

The game between Airbus and Boeing regarding developing UHCA, discussed in Case Study 2.2, can be seen as a chicken game in strategic form. Both manufacturers had two options–develop the UHCA and not develop it.

In 'First Mover's Advantage' section of Chapter 2, we discussed the scenario as a sequential move game. Indeed if the first mover developed the super jumbo the late mover should restrain. But before anyone made any move the scenario could be seen as a simultaneous move game. Using the same payoffs as before, we can represent the simultaneous move game in strategic form as the payoff matrix given in Figure 4.7.

There are two Nash equilibria—(develop, not develop) and (not develop, develop). Because it is a chicken game it makes

Figure 4.7

		Firm B	
		Develop	*Not develop*
Firm A	*Develop*	–5, –5	**7, 0**
	Not develop	**0, 7**	0, 0

strategic sense for the players to commit to developing the UHCA. Whoever makes a credible commitment wins the game. That explains why Boeing made a declaration in the media about their plans to develop the super-jumbo. Whether that is credible enough or not is another issue, which we will address in Chapter 5.

War of Attrition

War of attrition refers to a situation in animal conflict, wherein two males of the same specie fight for mates till one of them die or gets severely wounded (Smith, 1974). The game is an extension of the hawk–dove game. In the one-shot hawk–dove game given in Figure 4.5, there are four possible outcomes. The game is over if one player chooses 'dove' and concedes defeat, or if they both choose 'dove' to make peace. However if both the players choose 'hawk' the game continues, and it becomes a 'war of attrition'. The model of war of attrition can be used to analyse many human conflicts, including those in the domain of business. Attrition warfare is a well-known military strategy, conceptualized by ancient Chinese war strategist Sun Tzu (McNeilly, 2001). The idea is to wear down the enemy through repeated defeats in battles and thus enforce them to retreat. During the invasion of Iraq in 2003, the US military wanted to execute the same strategy. But a war of attrition can be very costly when both sides are equally strong and neither gives in. Let us now understand war of attrition using a business context.

Market Expansion as a War of Attrition

In certain industries, due to existence of increasing returns, along with limited demand, one dominant firm captures a very large share of the market. Such markets are generally called winner-takes-all markets. Market expansion game in these markets, before the consolidation, becomes a war of attrition. Typically, the firms need to make a very large initial investment to begin operations in such industries. In order to recover this investment the firms must sell enough. If the demand in the market is limited, firms become desperate to lock-in customers during the nascent stage of the industry. The lock-in is possible if using the product requires the consumer to invest in some devise or skill, which is typical of information and network industries (Arthur, 1996; Shapiro and Varian, 1999). Lock-in of consumers is also possible through creation of inertial brand loyalty (Demsetz, 1982; Wernerfelt, 1991). The firm that succeeds in locking-in the consumers wins the market, while the other firms are forced to exit. But till the relatively weaker firms quit and the market gets concentrated, all the firms lose money. Case Study 4.2 discusses a similar situation in the industry of direct-to-home TV broadcasting service in the UK during the 1990s. Though the case is quite old, it is a perfect example of war of attrition in business played between two equally powerful firms. A business situation between two players qualifies as a war of attrition if both lose money over a period in hope of monopolizing the market by imposing exit on the other. Since 2013, Costa Coffee and Starbucks are expanding rapidly in China. Starbucks started operations in China from 1999 and till 2012 they opened about 550 outlets, mostly in the big cities like Shanghai, Beijing and Guangzhou. But between 2012 and 2014 they added close to 1,000 outlets. Costa Coffee started China operations in 2006 and opened only 100 shops in the first six years. In 2014, they had 25 per cent of the market share, and have plans to have 2,500 outlets by 2018. The coffee shops are not exactly bustling in China, which is primarily a tea-drinking country. With more spacious coffee shops and less customers it may seem that the coffee giants are losing money in China. If that was the case, we could have analysed the situation as a war of attrition. But it is not. They are not losing money because they charge more in China. Starbucks charges about 20 per cent more in China, compared to USA, on coffee and espresso products. Both Starbucks and Costa are attracting Chinese customers by offering different varieties of tea and local specialty snacks. They are not just increasing the number of outlets, but also growing in sales revenue. So, it is an expansion war, but not really a war of attrition where the contestants bleed severely. On the other hand, price wars too do not qualify as a war of attrition game. A price war if often a Prisoner's Dilemma, as discussed in 'Grab the Deal—Discount Offer on Price as Prisoner's Dilemma' section of the previous chapter.

Case Study 4.2: War of Attrition in Satellite Television Market of UK

The technology of direct broadcast satellite television or DBSTV was innovated in the 1980s to improve quality of television broadcasts. Like the existing satellite television services, this technology also uses geostationary communication satellites to transmit and receive signals using satellite dishes. But the DBSTV system required much smaller dishes. The technology reduced dish size from 1.2 metres to 60–80 cm. This innovation made dish installation possible even in apartment buildings and congested habitats.

The broadcaster needs to invest in satellites, uplink satellite dishes and receiving dishes. Each subscriber household should have an exclusive receiving dish connected to the TV sets, through a set-top box interface. Each TV set requires a set-top box. The financial implication of this technology is high fixed cost. Faster recovery of the investment on satellites and uplink facilities require fast expansion. The viewers, on the other hand, could be reluctant to switch from cable TV to satellite TV because of higher installation charges. However, once a subscriber installs the dish and set-top box, making a fairly high investment, she gets locked-in due to the sunk cost nature of the installation cost incurred by the consumer. If a subscriber wants to switch, she cannot recover her investment on initial installation. This makes market consolidation inevitable. If the competing service providers foresee a possibility of monopolization due to limited demand, a war of attrition is likely to take off. That was exactly what happened during a period of six months starting April 1990.

(Case Study contd.)

(Case Study contd.)

British Satellite Broadcasting (BSB), a consortium of five companies, was formed in 1986 and they won 15-year license to operate three channels using DBS system in the UK. Back in 1977 International Telecommunication Union adopted an international plan under which each country was allotted specific frequencies for domestic broadcasting services. BSB obtained the sole right to broadcast satellite TV on frequencies allotted to UK. Armed with the monopoly right to provide satellite television in the UK, BSB invested £200 million to buy two satellites. They calculated that to break-even at somewhere around 3.5 million installations, which they expected to achieve in the fourth year from launch, they needed to price installations at £250. They planned to go on air in 1989 and expected to install 400,000 dishes in the first year. They planned to install another 2 million by 1992, and expected to break-even by 1993–94. Hoping to bring down installation charges thereafter, BSB planned 6 million installation by 1995 and 10 million by 2001.

The plans went for a toss when Rupert Murdoch, who failed to obtain DBS license, expanded Sky Channel and created Sky Television Network—a four channel UK based satellite TV service. This was made possible by Luxembourg based Société Européenne des Satellites (SES) who launched Astra 1A satellite on 11 December 1988. Astra was the first medium power satellite in Western Europe that was not part of ITU band. It transmitted signals in K_u band and made reception possible with 90 cm dishes. Hence the Astra based service became a very close substitute of the DBS system. Since they used Astra satellites Sky Television had no major investment. Their total start-up cost was £100 million. Their expansion plan was far more aggressive compared to BSB. Sky TV planned to install 1 million dishes by 1989 and 5–6 million by 1994. They expected break-even by 1991–92 at 3–4 million installations. In response to Sky's entry announcement BSB revised expansion plan to 5 million by 1993 and 10 million by 1998. They needed to break-even before Sky did. BSB and Sky both had movie channels and wanted to acquire exclusive

(Case Study contd.)

(Case Study contd.)

British television rights for Hollywood films. Even before any of them went on air, by end of 1988 they have spent a total of £670 million in bidding for films.

BSB went into financing troubles and couldn't launch in 1989. Sky went on air in 1989 and in the first year installed 600,000 dishes, 400,000 short of their target. The demand was not great. BSB went on air in April 1990. Between April and July 1990 BSB installed 25,000 dishes. During the same period Sky installed 292,000. In a desperate attempt to win the race they slashed installation charges, and in the period August–October 1990 BSB installed 125,000 dishes as against Sky's 58,000. It was a desperate scramble and both the service providers were losing money. At the end of October 1990, BSB was losing £6 million to £7 million a week. Sky too was losing £2 million a week. At that point in time BSB had 175,000 installations vis-à-vis 950,000 of Sky. Both the companies were having trouble financing this war of attrition. On 2 November 1990, they merged and formed British Sky Broadcasting (BSkyB) with a 50:50 ownership between BSB's shareholders and Sky TV's parent News Corporation.

Source: Ghemawat, 1997.

The contest between the BSB and Sky TV Network was a war of attrition. After Sky declared their expansion plan, BSB revised their plan and wanted to expedite expansion. Clearly, they wanted to lock in as many subscribers as possible in a short span of time to beat Sky. After they lost out in the first leg during the period April–July of 1990, BSB accelerated expansion in the next quarter. Indeed they didn't want to give away the market to Sky. They lost money in the process, but also made Sky lose money. That is the hallmark of a war of attrition. To model a war of attrition, consider a situation where two firms compete for a monopoly position in the market. Till one of them quit they play a hawk–dove game period after period. Suppose, to stay in the market a firm must sink in an amount x each quarter till the war is over. If one of the firms quit in some quarter the war is over. The other firm

Figure 4.8

Firm B

		Stay	Quit
		Stay	*Quit*
Firm A	*Stay*	$-x, -x$	**9x, 0**
	Quit	**0, 9x**	**5x, 5x**

gains monopoly position in the market. Let the monopoly profit earned in the market over a time period in future be $10x$. If both firms quit, they merge. The merged organization becomes the monopoly, and the monopoly profit is shared between the owners of the two firms. In any quarter, the game in strategic form is given by the payoff matrix in Figure 4.8

If one firm stays and the other quits, the firm that stays become monopoly and gets $10x - x = 9x$. However, if both quit and merge, they don't spend resources in fighting the war. For a one-shot game, like the one shown in Figure 4.8, there are two Nash equilibria—(stay, quit) and (quit, stay). But the (stay, stay) outcome is very likely. If the outcome in any quarter is (stay, stay), the game is again played in the next quarter.

Knowing that the game will continue if the outcome is (stay, stay), at the inception of the war the players must decide how many quarters they will stay. If we ignore the time value of money it may seem that the players should not stay for more than 10 quarters, as after 10 quarters even if a firm wins the monopoly position in the market they have already spent more than the monopoly profit they could earn in future. This argument is flawed. The money spent in waging the war is sunk cost. If Firm A quits in the 11th quarter, the best response for Firm B is to stay in the 11th quarter. That way Firm B recovers $10x$ after spending $11x$, whereas Firm A recovers nothing of the $10x$ they spent. The argument is true for both the firms. But if both stays the war continues, and that way the war may continue forever. Merging and sharing the monopoly profit may be a better option. Since after merger each owner gets $5x$, they should merge within the first five quarters. But if Firm A quits and proposes the merger, in any quarter the best response for Firm B is to stay. Suppose Firm A quits and proposes the merger in the fifth quarter. By then each firm have lost $4x$ in the previous four quarters. By agreeing to the merger Firm B is left with x. If they stay they get $10x - 5x = 5x$. So Firm B will stay, and hence Firm A too will stay resulting in continuation of the war.

If at all any firm must decide to quit, they must do so in the first quarter. In that case the war would have been avoided. But if a firm stays, it should stay for one more quarter than the rival. Suppose Firm A decides to stay till Firm B quits. In that case Firm B's best response is to quit in the first quarter instead of losing money by staying for some quarters. Similarly if Firm B decides to stay till Firm A quits, the best response of Firm A is to quit in the first quarter. This essentially means that there are two Nash equilibria in the war of attrition game—one of the firms quits in the first quarter and the other stays. The concept of Nash equilibrium fails to theorize the rationale for war of attrition to continue. However, we can intuitively understand why the war might continue. Both firms wait for the other to quit and the war of attrition continues. Indeed it cannot continue forever. One of the firms needs to convince the other firm that they are committed to stay till the end. To see how that can happen we will revisit the game in Chapter 7.

In this chapter, we discussed a few classes of simultaneous move games in strategic form. We identified the Nash equilibria in these games, but didn't discuss the conflict resolution for the games with multiple Nash equilibria. In the next chapter, we will discuss strategic moves that might be useful in resolving the conflict. The point is to out-think the rival.

5

Strategic Moves: Threats, Promises and Commitment

In simultaneous move games unique solutions may not exist. Particularly in coordination games and anti-coordination games, where there exist multiple Nash equilibria, the concept of Nash equilibrium is not very useful in practical decision-making. We saw in Chapter 4 that in such games, the players may not be able to reach Nash equilibrium. Changing the order of moves from simultaneous to sequential might help the players reach Nash equilibrium. If the rules of a game are not fixed, self-interested players will want to manipulate the rules to their own advantage. Such manipulative manoeuvring of the game is referred to as strategic moves. In this chapter we will discuss strategic moves that not only are useful in resolving conflicts, but also help the players to win a game.

Changing the Order of Moves

Let us revisit the BoS game. While introducing the game in the section 'Coordination Games' of Chapter 4, we considered the game to be a simultaneous move game as the players could not communicate with each other. They were mad at each other and hence weren't picking up each other's calls. In order to change the game from a simultaneous move to a sequential move, one of the players may send a text message. Such a move will be an example of strategic move. Let us see how such a move might be helpful. Suppose the woman sends a text message. This means in the sequential move game the woman makes the first move. She might either inform her partner that she is going to go for her preferred entertainment, that is, football match, or she might inform him that she is being nice and choosing to go for the concert, which is her partner's preferred entertainment. Knowing the decision of his partner the man can choose his best response. The game tree is given in Figure 5.1.

The woman should be able to foresee that if she texts that she decided to watch the football match her partner will follow suit, and if she texts that she is going to the concert then her partner will happily go to the concert. Since her payoff is more when they watch the football match together than when they together go to the concert, she should text that she is going to watch the football match. We are applying the method of backward induction, which was discussed in Chapter 2. The best responses are indicated by the bold branches.

We can see that the text message serves two purposes. Firstly, the one-way communication helps them avoid coordination

Figure 5.1

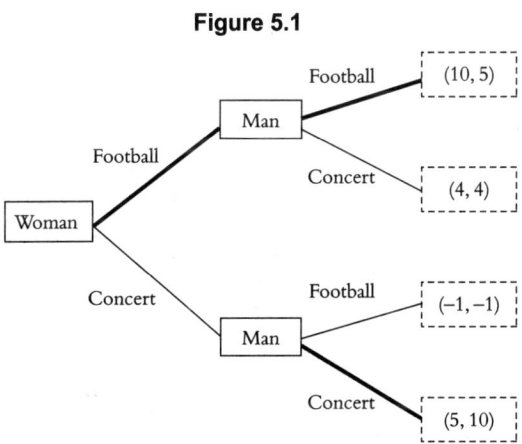

Figure 5.2

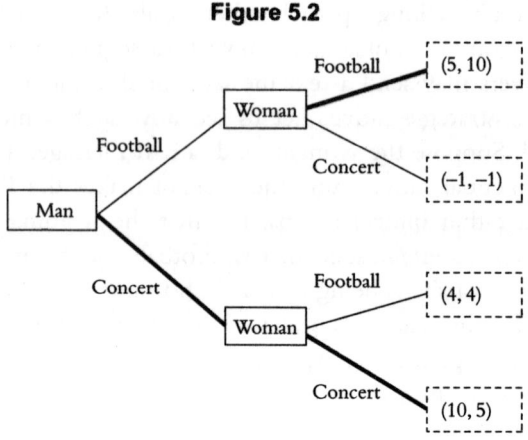

failure. Secondly, whoever sends the text message first gets the first mover's advantage. If the man sends the text message, first he will let his partner know that he is going to the concert, which will in turn enforce the woman to go to the concert. The game tree with the man moving first is shown in Figure 5.2. Here the first payoff is the man's payoff. The bold branches indicate the best responses.

Gaining Credibility through Commitment

Since there is first mover's advantage in the sequential move BoS, both would want to grab the opportunity and send the text message before the other. If sending a text message was enough, then it would have boiled down to a fastest finger contest. Is a text message credible enough? Suppose the woman sends the text first. Is there any reason for the man to take it in face value and follow suit? What if he puts his foot down and sends a reply like the following? *Go wherever you want. I'm going to the concert. The passes are attached.* Note that if the man goes to the concert, the woman's best response is to follow suit! The passes are proof of his determination. The fact that he spent the money for buying two passes or put in effort to get hold of the passes makes his message more credible. Just saying that he obtained the passes wouldn't have been credible. It is important to exhibit his credibility, which he did by attaching the passes.

The system adoption game between two business partners, discussed in 'Technology Adoption in Presence of Network Effect—A Coordination Game' section of the last chapter, is also a BoS game. Either A should switch from Windows to Mac or B should switch from Mac to Windows. There are two Nash equilibria of the simultaneous move game, as shown in the payoff matrix given in Figure 4.3. If A does not switch, the best response of B is to switch, and vice versa. Since there is a switching cost no one wants to switch but wants the other to switch. In this game too, the problem of coordination failure could be avoided by changing the sequence of moves. The game tree with A deciding first is shown in Figure 5.3.

If A says that she will not switch it induces B to switch. On the other hand, if she switches, B won't switch. Thus coordination failure is avoided. Since there is a switching cost, A will want B to switch. There is first mover's advantage here too. The partner moving first will want to induce the other to switch. However, to be credible, the first mover must commit herself or himself to not switching. The game tree with A moving first and committing not to switch is shown in Figure 5.4.

Figure 5.3

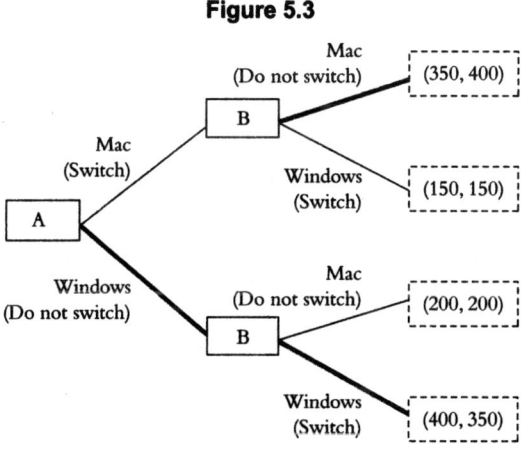

Figure 5.4

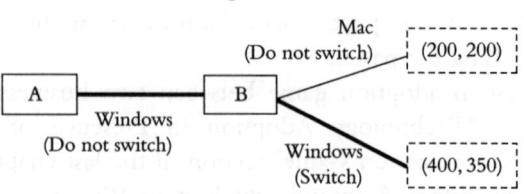

With the commitment, switching to Mac is not an option for A. Hence that branch of the game tree was removed. Such a commitment might be made by some investment in the existing system. For example, A can buy some Windows compatible software, or some PC compatible hardware, or might upgrade her versions of Windows and MS Office. The effect is same as buying the concert passes in the context of the sequential move BoS game discussed earlier in this section.

Making Commitment by Elimination of Options

In anti-coordination games, players can make commitment by eliminating some of the options available to them. To understand the role of commitment in anti-coordination games discussed in 'Anti-coordination Games' section of the last chapter, let's revisit the chicken game. In the sequential move chicken game too there is first mover's advantage. Suppose driver 1 thumps his chest and asserts that he will not chicken out. If driver 2 takes that in face value and believes that driver 1 is not going to chicken out, his best response is to chicken. The game tree is given in Figure 5.5.

Is it credible that driver 1 won't chicken? It isn't. If driver 2 dares to go straight onto driver 1, despite his arrogant assertion driver 1 may flinch and chicken. He didn't do anything that makes him committed to his assertion. But suppose he jams his steering wheel or even pulls it out, that very action makes him committed to dare. By removing the steering wheel driver 1 is eliminating his

Figure 5.5

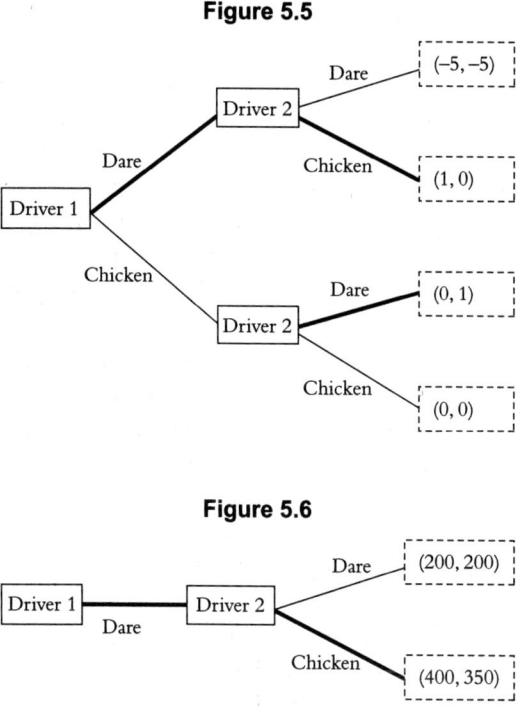

Figure 5.6

option to chicken out. Figure 5.6 shows the game tree with the option to chicken removed.

Since driver 1 does not have the option to swerve, his assertion of not chickening out becomes credible. To gain credibility, we need to reduce our strategy space (Schelling, 1966, 1980).

Boeing Failed to be Credible

Recall the Case Study 2.2 from Chapter 2. In the game of developing the ultra-high capacity carrier between Boeing and Airbus, there was a first mover's advantage. Whichever firm would have moved first and developed the super-jumbo jet would have pre-empted the other from developing a similar aircraft. The strategic form of the game was discussed in 'Anti-coordination Games' section of

Chapter 4, and we understood that there existed two Nash equilibria. In one of the equilibria, Airbus develops the super-jumbo and Boeing does not, and the other way round in the other equilibrium. The problem of multiple equilibria could be avoided in a sequential move game. The first mover gets the advantage if they can credibly convince the rival that they are committed to developing the super-jumbo. In an attempt to do exactly that, Boeing declared in *Business Week* that they were developing a carrier of capacity 600 to 800 seats, which they claimed to be the "biggest and most expensive airliner ever." But that public declaration on mass media was not enough to discourage Airbus from going ahead with their plan of developing A380. What went wrong? Boeing lacked credibility. They had no commitment to back their claim. On the contrary, their credibility was dampened by the fact that they had B747-400, and if they developed the super-jumbo that would have cannibalized it. In fact *Business Week* only reported that "Some in the industry suggest Boeing's move is a bluff to pre-empt Airbus from forging ahead with a similar plane."

What Boeing could have done to gain credibility? There were various possibilities. They could have made some credible and sizeable investment. They could have possibly gotten into a R&D joint venture with one of the engine manufacturers like Rolls Royce or Pratt & Whitney. Such a move would have been very credible because it would not only require them to make a sizeable investment, but shelving the plan to develop the super-jumbo would have jeopardized their long-term relation with their engine supplier. Stopping production of 747s would have made them more credible by reducing their product line. In absence of B747 products, Boeing would have been more committed towards development of the super-jumbo.

Sinking Ships or Burning Bridges—Military Strategy

Reducing options to pre-empt retreat is a military practice. Hernán Cortés de Monroy y Pizarro, the Spanish conquistador, landed at Vera Cruz (a coastal town in Mexico) in April 1519. He commissioned the expedition against the will of Diego Velázquez, who was then the governor of Cuba. Knowing that he will be either killed or imprisoned on his return to Cuba, Cortés had no option but to keep fighting and conquering the Aztec empire. In July 1519, some of his men conspired to seize a ship and escape to Cuba. Cortés had only about 650 soldiers with him and could not afford losing any of them. In response to the mutiny, he decided to scuttle all his 11 ships. That made retreat impossible for his men, and forced them to fight and conquer the land. Understanding that the Spaniards were there to stay and conquer, the local Cempoalans surrendered to them and became allies in the Spanish conquest of the Aztec land.

The idea of pre-commitment is appealing in military strategy, and often referred to as burning the bridge strategy. To put it in form of a game let us consider a situation wherein there is a conflict between two countries over a disputed island. The island is connected to mainland of country B through a bridge over a channel. Currently the army of country B is occupying the island, but knows that country A is contemplating an attack on the island from the sea. Refer to the diagram given in Figure 5.7.

The situation can be modelled as a sequential move game. Country A must decide whether to attack or not. If country A does not attack, then country B retains the island. If country A

Figure 5.7

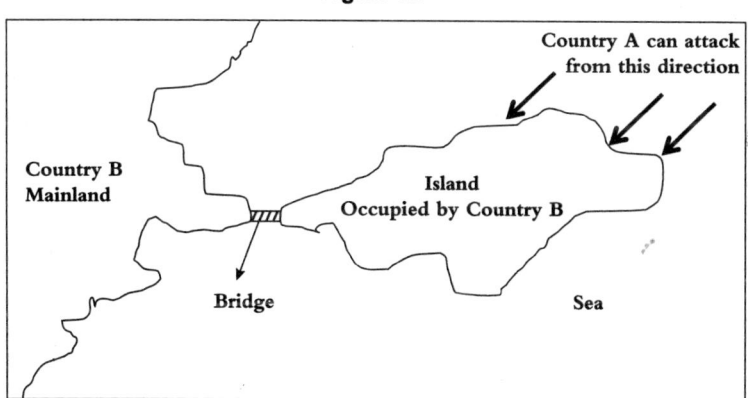

Figure 5.8

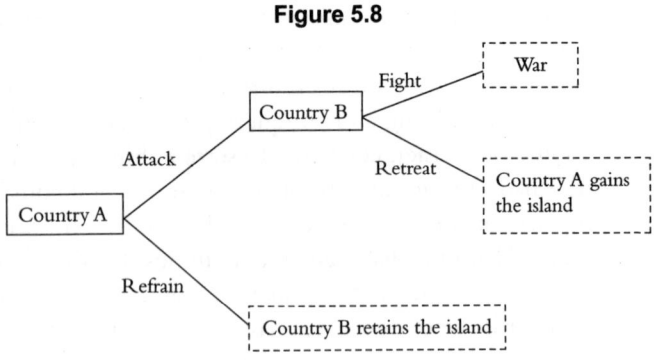

attacks, country B may either fight or retreat. If country B retreats then country A gains the island. If country B fights back then there is a war, which is the worst outcome for both countries. Figure 5.8 depicts the game in extensive form.

We need payoffs to solve the game using backward induction. Let the value of the island be 10 to both countries. Suppose cost of war is 6. Both countries understand that in case of a war there is only a 50 per cent chance of winning the war. With a 50 per cent chance of winning the expected gain from the war is only 5. This means that the cost of war exceeds the expected gain for both countries, and each country gets a net expected payoff of −1 in case of a war. Putting these payoffs in the extensive form representation of the game, the game tree looks like the one given in Figure 5.9. The bold branches indicate the optimal decisions at the respective nodes.

Country A should foresee that if they attack, the best response of country B is to retreat. So, country A will attack. Country B can change the outcome drastically by burning down the bridge connecting the island to their mainland. This may seem like madness. But sometimes it is rational to act irrational. Here burning the bridge is a pre-commitment to fight. In absence of the bridge, the army of country B cannot retreat. Eliminating the option to retreat country B commits her to fight. Knowing that an attack will invariably result in a war, the rational decision of country A will be to refrain from attacking the island.

Figure 5.9

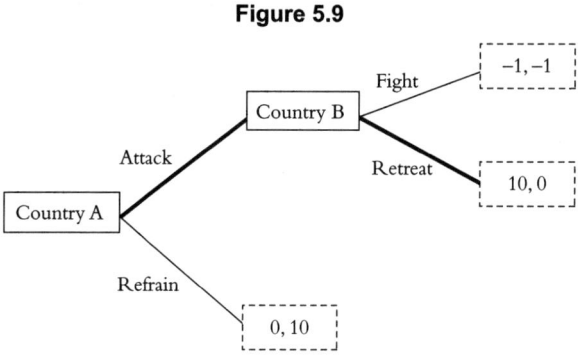

The decision to burn the bridge is a strategic move. It is a pre-emption strategy by country B. Incorporating the strategic move the game tree is given in Figure 5.10. In the game tree given in Figure 5.10 country B is the first mover, and hence the payoffs are given as (country B's payoff, country A's payoff). The best responses at different nodes are indicated by the bold branches. Foreseeing that burning the bridge will pre-empt country A from attacking the island, and that in presence of the bridge country A will attack the island forcing retreat on the army of country B,

Figure 5.10

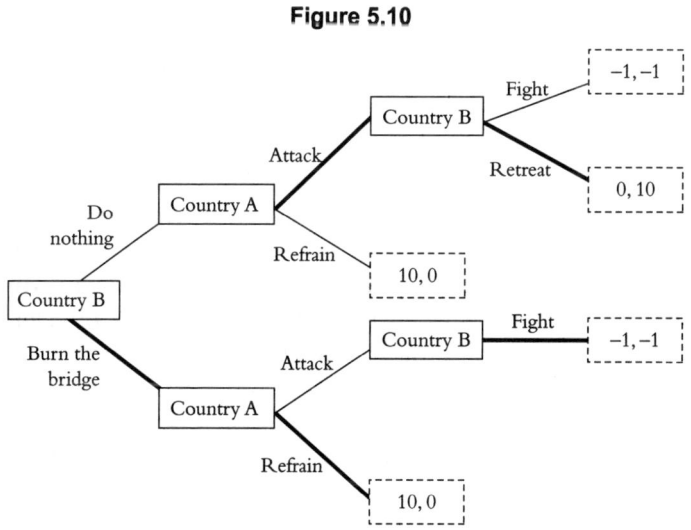

country B should burn the bridge. However irrational it sounds, it is the rational decision.

Strategic Use of Dominated Strategy

In this section, we will discuss how changing the order of moves helps in gaining advantage even in games where there exists a unique Nash equilibrium. Consider the following game between two asymmetric firms. The firms can either choose an 'aggressive' strategy or a 'passive' one. When the smaller firm adopts 'aggressive' strategy it hurts the larger firm. However, it is not in the best interest of the larger firm to retaliate aggressively. For example, if the firms compete in prices in a particular product category, aggressive strategy is price cut. If the small firm cuts price, the large firm loses market share. In retaliation, the large firm may cut price and that may trigger a price war. When both firms cut prices and engage in a price war, both firms lose money. But the fall in price impacts the bottom line of the large firm more severely than the small firm. Hence, engaging in a price war is not in the best interest of the large firm. The generic game in strategic form is given in Figure 5.11. The payoffs may be taken as the firms' contributions, in crores of rupees per month.

When both firms choose 'passive', the payoffs of both firms are more than their respective payoffs vis-à-vis when both choose 'aggressive'. But when one firm chooses 'aggressive', it is comparatively better-off than the rival who chose 'passive'. Both choosing 'aggressive' is the worst scenario for both firms. For the large firm, the dominant strategy is 'passive'. But if the large firm chooses 'passive', the small firm is better-off choosing 'aggressive'. The only Nash Equilibrium of the game is (passive, aggressive) which leaves the large firm with a payoff of 15. The choice of 'aggressive' strategy by the small firm hurts the large firm, but the large firm cannot retaliate by choosing 'aggressive' because that leaves it with a payoff of 12. The large firm is best off when both firms are 'passive'. But the Nash equilibrium payoff is not even the second

Figure 5.11

Small firm

		Aggressive	Passive
Large firm	**Aggressive**	12, 4	17, 5
	Passive	**15, 7**	18, 6

best for the large firm. In the simultaneous move game, the large firm is forced to accept the third best payoff among the four possibilities. Can it do better?

In a sequential move game, if the large firm makes the first move and chooses 'aggressive' it induces the small firm to choose 'passive'. The extensive form representation of the sequential move game is given in Figure 5.12.

The optimal decisions at different nodes are shown by the bold branches in the game tree. If the large firm chooses 'aggressive' it induces the small firm to choose 'passive'. But if the large firm chooses 'passive' the small firm will choose 'aggressive'. Foreseeing that, the large firm should choose 'aggressive' and thus can force the

Figure 5.12

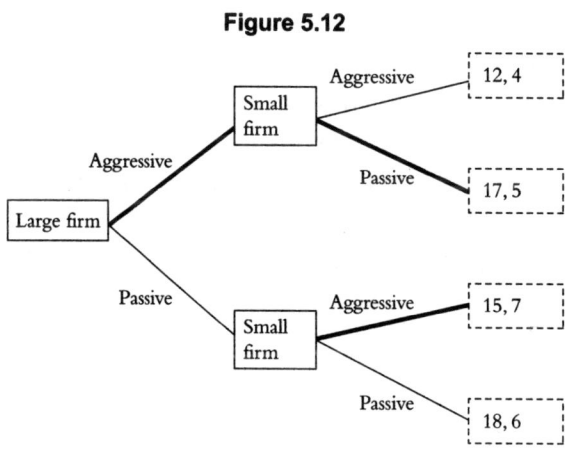

small firm to choose 'passive'. So, by changing the order of moves from simultaneous to sequential, and by moving before the small firm, the large firm can increase its payoff from 15 to 17.

A similar situation was encountered by Hindustan Lever Ltd (HLL, now Hindustan Unilever—HUL) during the 1980s when Nirma's aggressive pricing pushed them off the perch in the Indian detergent market.

Case Study 5.1: Hindustan Lever Ltd and Nirma

In 1959, HLL introduced the Surf detergent powder to the Indian consumers, who were primarily users of detergent bars. Surf became a huge success in next decade, though it was confined to the urban market and was affordable only to the upper-middle class. Nirma was introduced in 1969, but remained confined to the Gujarat market. Priced at ₹3 per kg, Nirma's presence soon extended to the whole of Western and Northern India. In the year 1977, Nirma had 11.9 per cent market share, vis-à-vis 30.6 per cent of Surf. Nirma was priced at ₹4.5 per kg. HLL did not reduce the price in response to Nirma's entry. In fact HLL increased the price of Surf from ₹10.15 per kg in 1976 to ₹12.8 per kg. As a result, between 1976 and 1977 Surf's market share reduced by 19.8 per cent. By 1984, Nirma became the most selling brand leaving Surf behind, and by 1987 Nirma had 61.6 per cent market share vis-à-vis 7.4 per cent of Surf. In the year 1987, Surf was priced at ₹27.10 per kg, whereas Nirma was priced at ₹8.5 per kg. At that point in time in 1987 HLL introduced low-priced detergent Wheel to counter Nirma.

Source: Case Study 'Hindustan Lever Limited: Levers for Change' by Charlotte Butler and Sumantra Ghoshal taken from Ghoshal et al. (2002).

Till 1987, HLL chose to be passive. They did not respond to Nirma's pricing by cutting price. They thought Nirma's price was unsustainable. Also they didn't want to dilute the brand value of Surf, which was a premium brand. Introduction of

Wheel was the aggressive strategy. By the turn of the century, Wheel became the highest selling brand in the low-priced detergent category replacing Nirma.

Strategic Move vis-à-vis Incentives

In coordination games, it seems that incentivizing the partner might be helpful. For example, consider the RFID technology upgrading game between the supplier and retailer discussed in 'Technology Adoption in Presence of Network Effect—A Coordination Game' section of the last chapter. A scrutiny of the payoffs in Figure 4.4 reveals that a transfer of payoff amounting just more than €0.12, from the retailer to supplier, when both switch to case-tagging system, makes adoption of case tagging the focal equilibrium of the game. When they adopt the case-tagging system, the supplier bears a re-training cost of €0.05 per pallet and average fixed cost of €0.1 per pallet. If the retailer bears this training cost and fixed cost of implementing the new system, it amounts to a transfer of €0.15 from the retailer to the supplier. The modified payoff matrix is given in Figure 5.13. All payoffs are same as in Figure 4.4 except for under the outcome wherein both switch to case tagging.

Even with the modified payoffs there are two Nash equilibria—one where they continue with pallet tagging and the other where they switch to case tagging. Still there is no dominant strategy for any player. However, the payoffs are higher for both players when

Figure 5.13

		Retailer	
		Continue with pallet-level RFID system	*Switch to case-level RFID system*
Supplier	*Continue pallet tagging*	**0.45, 0.10**	0.15, –0.36
	Switch to case tagging	–1.27, –0.06	**0.48, 4.89**

they switch to case tagging, that is, the transfer of payoffs made both switching to case tagging the focal equilibrium of the game. As discussed while analysing the assurance game in Chapter 3, rational players will be able to coordinate on the focal equilibrium in presence of pre-play communication.

This example shows how incentivizing one player by transferring payoff from the other might be helpful in arriving at the desired equilibrium, which in this case was upgrading the technology and adoption of the case-tagging system. This was possible because the gain from upgrading the technology was large enough for the retailer. Nevertheless, the retailer would have preferred a solution wherein the case-tagging system would have been adopted without transfer of payoff from the retailer to the supplier.

Since the pallets with RFID tags that are sent by the suppliers should be scanned when it arrives at the retailer's DC, the game may be perceived as a sequential move game where the supplier moves first. The game tree with the supplier moving first is shown in Figure 5.14. Here we are not considering transfer of payoffs, and the payoffs are same as those given in 'Technology Adoption in Presence of Network Effect—A Coordination Game' section of Chapter 4.

Figure 5.14

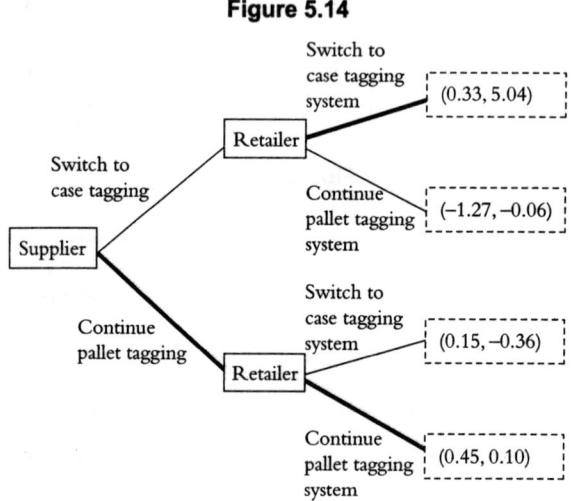

The supplier can foresee that if they continue with pallet-tagging using old generation RFID tags, the retailer won't be able to switch to case-tagging system. Hence they will continue with the existing technology, halting adoption of new technology. The equilibrium path of the sequential move game is shown with bolded branches of the game tree in Figure 5.14.

Now suppose the retailer moves first and decides whether to replace the old system with new system that reads RFID tags only if the cases are tagged with new generation tags, or to continue with the old system, that is, the retailer makes the first move and decides whether to switch to the case-tagging system or to continue with the old pallet-tagging system. The game tree with the retailer moving first is shown in Figure 5.15.

Since the retailer is the first mover, now the retailer's payoffs are written first in Figure 5.15. With this modification we see that if the retailer continues with the old pallet-tagging system the supplier will continue to tag only the pallets. But if the retailer replaces their existing systems with new systems that read tags only when cases are tagged with new generation RFID tags, the suppliers will switch to case tagging using the new generation tags. Here the retailer does not need to transfer payoffs to the supplier.

Figure 5.15

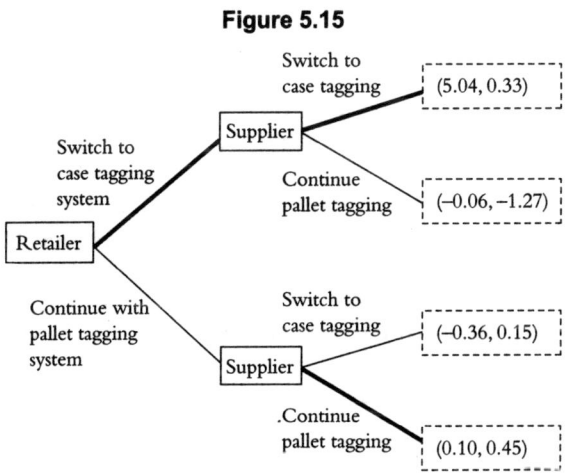

Strategic Location Choice

Choosing the location of your retail outlet may be a strategic move. To illustrate the point let us take the example of two coconut-water vendors on a beach front. Imagine a 1-km stretch on a beach where two coconut-water vendors, A and B, choose their locations. Assume that the density of beachgoers is uniform along the stretch of the beach. Since the beachgoers don't perceive any difference in quality of coconuts sold by A and B, if the prices are same they drink from the vendor who is closest to them. Where will A and B position themselves on the beachfront? The problem of the coconut-water vendors is not very different from the location choice game played by Starbucks and Costa Coffee in the central business districts of Beijing and Shanghai.

Case Study 5.2: Starbucks and Costa in China

In Beijing's Blue Harbour International Business District, there is Starbucks on one of the outer street corners, and Costa in the central courtyard. While local Chinese coffee shops dread the opening of Starbucks in their neighbourhood, the British chain shop goes out of its way to try to be as close to as many Starbucks as possible. Almost all its shops are right beside Starbucks.

Since 2013, Costa Coffee and Starbucks are expanding rapidly in China. Starbucks started operations in China from 1999 and till 2012 they opened about 550 outlets, mostly in the big cities like Shanghai, Beijing and Guangzhou. But between 2012 and 2014 they added close to 1,000 outlets. Costa Coffee started China operations in 2006 and opened only 100 shops in the first six years. In 2014, they had 25 per cent of the market share, and have plans to have 2,500 outlets by 2018.

Source: Author.

Suppose the vendor A first chooses the location on the beachfront. He can choose any location. The beachgoers decide whether of have coconut water depending on price and their distance from the vendor. A longer distance that a beachgoer might need to walk to fetch the coconut water from the vendor may be compensated by a lower price. If the vendor A positions himself at one of the ends of the 1-km stretch, the beachgoers at the other end of the stretch might not come to drink coconut water from him unless the price is sufficiently low. Instead if he chooses position at the midway, every beachgoer is within 500 metres from him. This location increases his demand at any given price. Now if vendor A chooses to locate himself at the midway, where should vendor B position himself?

Vendor A positioned himself at the 0.5 km mark, as given in Figure 5.16. If vendor B positions himself at the far end, that is, at the 1-km mark and sells coconut water at the same price as vendor A, then all the beachgoers between the 0-km mark and 0.75-km mark will go to vendor A and the ones between 0.75-km mark and 1-km mark will go to B as only the beachgoers located between the 0.75-km mark and 1-km mark find vendor B closer to them. Refer to Figure 5.16. Only way the vendor B can increase market share is by reducing price, or by offering some value addition. Since the product is a commodity, it is impossible to differentiate in quality. So the vendor B will be forced to reduce price. However, if vendor B moves closer to the centre, he can increase his market share. Consider the locations given in Figure 5.17 when vendor B positions himself at the 0.7-km mark.

Now all beachgoers between 0.7-km mark and 1-km mark become the captive market for vendor B. The beachgoers between

Figure 5.16

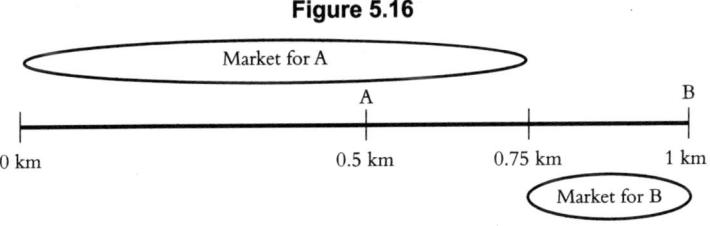

Figure 5.17

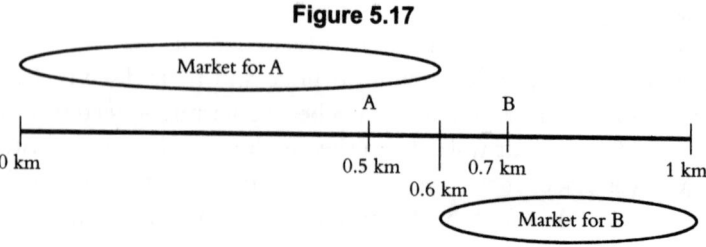

0 km and 0.5 km remain captive to vendor A. The beachgoers between 0.5 km and 0.7 km gets split with the ones between 0.5 km and 0.6 km going to A and the ones between 0.6 km and 0.7 km going to B. Thus by changing the location, without reducing price, vendor B could increase the market share from 25 per cent to 40 per cent. If vendor B positioned himself just next to vendor A, he could have increased his market share to 50 per cent, and that is the best he could do. When the vendors are just next to each other none of them get any locational advantage. However, moving beyond the midpoint will again reduce the market share of vendor B. If the products are identical then the market gets equally shared.

What Costa Coffee is doing in China is not very different from what vendor B could do. Being in close proximity to Starbucks they are not letting Starbucks get any locational advantage. Since the café-going Chinese consumers are not very prices sensitive, but rather brand conscious, the competition between Costa Coffee and Starbucks zeroes down to a branding war resulting in improved quality and service.

So far in this book we have discussed games where the players interacted once. The dynamics of strategic interactions change drastically if the players interact repeatedly. In the next chapter, we will discuss games with repeated interactions.

6

Trust, Credibility and Collusion in Repeated Games

In Chapter 3, we came across simultaneous move games where the players are forced to accept mutually damaging outcome. All such games can be classified as prisoner's dilemma. In prisoner's dilemma class of games, the Nash equilibrium is inefficient as there exist outcomes within the games wherein all players could have been better off. In the classic prisoner's dilemma example, prisoners choose to confess due to the lack of trust. However, had they both not confessed, each would have spent less years in prison. Our logical quest should be to find a way to reach the win–win outcome that exists within the game. In order to be able to do that, players need to overcome the incentive to cheat and trust each other. To be precise, in a prisoner's dilemma class of game, players must choose to play their dominated strategies instead of playing the dominant strategies. That is irrational in one-shot games, but works fine when the game is repeated for a large number of times. In this chapter we will see how optimal mix of trust, threat and optimism can bring in cooperation and sustain mutually beneficial outcome when the game is played repeatedly.

Repeated Games

By definition, a one-shot game is a simultaneous move game where players cannot do anything that affects the other after the game is over. When games are one shot in nature, it is impossible to use threats or incentives conditional to the choices made by the other. Suppose, for example, in the original prisoner's dilemma problem, the prisoners had to make the pact without any threat against the breach of the pact. In order to ensure a binding commitment towards the pact, the prisoners should be able to pose the credible threat of punishment that is more damning than even 10 years in prison. For example, suppose the prisoners threaten each other that if one breaches the pact, the other who gets sentenced for 10 years will kill the traitor on his return from prison. Indeed, the outcome of being murdered is way more damning than spending 10 years in prison. But the threat is not credible. Ten years is way too long in the future and provides the opportunity to escape to the traitor. In a one-shot game, there is no trust between the players because of the absence of executable and credible threat.

Typically, market games are not one shot in nature. Firms are there in the market and they play the same games repeatedly over and over again. Since there is future, it is possible to ensure win–win outcome by the mix of cooperation and credible threat. The players need to make a pact or must have an understanding that ensures the win–win outcome. In order to sustain that cooperative win–win outcome, they must overcome their incentives to cheat over the other. So, the players need to use a credible threat of punishing the others if they cheat. In a repeated game, the players can trust each other because they know that they have ways to punish the others if they cheat, by defecting from the pact. The executable and credible threat helps in developing trust. Before we get down to the formulation of business strategies by the use of such credible threat in repeated encounters, let us go through a few behavioural real life examples from the human and animal world.

Grudger Birds

In his book, *Selfish Gene*, Richard Dawkins talks about a specie of bird that requires mutual preening. These birds get parasite ticks on their feathers. These ticks are vectors of some deadly diseases and the birds often get infected because of the ticks. For their survival, it is important to clean their feathers and the birds can preen most of their bodies using their bills. But there are certain parts of the body that they cannot reach with their bills. For example, to preen the top of their head or the back of their neck, they need the help of another bird. Mutual preening means helping each other in preening. It was observed that some birds are very clever. They get preened by another bird and before returning the favour they fly off. Dawkins refers to these birds as 'cheats'. The other type of birds that go around helping others in preening irrespective of whether they get preened in return are called 'suckers' by Dawkins. If these were the only two types of birds within the specie then the specie would have been extinct. 'Cheats' don't return the favour. So, a 'sucker' gets preened only if it meets another 'sucker'. Life is good for the 'cheats' as they get preened without spending energy in preening others. In a way, we can say that the payoff of a 'cheat' is more than that of a 'sucker'. So, the evolutionary process will favour the genes that carry the phenotype of 'cheat' behaviour. Over generations the number of 'cheats' will outweigh the number of 'suckers'. As the birds don't get preened when they meet a 'cheat', they will rarely get preened in a population wherein the majority are 'cheats'. So the birds get infected easily and over some generations the specie might become extinct. This process of extinction was confirmed by a computer simulation.

So, how is the specie surviving? The birds, who Dawkins thought of as 'suckers', are actually the smartest ones—the 'grudgers'. They behave strategically. They are fundamentally nice and begin with cooperating in what they understand as mutual preening. They preen the head and back of neck of another bird with whom they intend to develop a relationship of mutual preening. If the other bird returns the favour, they actually develop a long-term relationship of mutual preening. But if the other bird turns out to be a 'cheat', the 'grudger' remembers and never preens that 'cheat' ever in future. This behaviour is hardwired in the 'grudgers' and determined by a particular phenotype embodied within a genotype that is present in them. It is natural for them, but they are able to outsmart the other birds. In a population with a critical minimum proportion of 'grudgers', the 'grudgers' do best. The 'cheats' may get away with their behaviour a few times, but since preening is not a one-off affair, they will soon fail to find birds to preen them. As a result, they get infected and die. Since the 'cheat' phenotype does not pay well in a population with a sufficiently large number of 'grudgers', the phenotype gets dominated. The existence of the 'grudger' phenotype is critical for the survival of the specie.

The 'grudger' birds play a repeated game using a credible threat of executing a punishment on other birds that defect. Indeed, the 'grudger' birds are not strategic but hardwired by nature to play the repeated game in this manner. In the next section, we will discuss trigger strategies that are used by strategic players for playing repeated games. The nature of the trigger strategy is the same as what the 'grudger' birds do—using a threat of punishment to defectors in order to ensure cooperation.

Love Thy Neighbour as Thyself

That was the second commandment as given in Chapter 22 of the Gospel of Matthew (Matthew 22: 39). The same is said in various other forms in different religious texts. In general, in human society, neighbours maintain a cordial relationship and help each other in time of need. The kind of cooperation that is observed between neighbours is generally not seen between strangers. Indeed, neighbours do fight sometimes, but that is when the conflict becomes a constant sum game, where one's gain is the other's loss. If human beings were all as altruistic as the Samaritan in the 'Parable of the

Good Samaritan', they would have helped even strangers. Humans help neighbours because they need the reciprocal help in their own time of need. This reciprocal altruism works very much in the same way as the mutual preening by birds. Another example of reciprocal altruism among neighbours was discovered in marine life by famous sociobiologist R. L. Trivers. In his article, 'The Evolution of Reciprocal Altruism', Trivers provides the example of the synergic relationship between cleaner fishes and sharks within bounded regions under water. The underwater regions enclosed by coral walls or underwater mountains function pretty much the same way as the gated communities function in human society. The ecosystem within the bounded region is self-sufficient. Bigger fishes live on smaller fishes. Small cleaner fishes and shrimps live on the algae and other microgreens that grow on the bodies of the sharks. The sharks can identify these cleaner fishes by their colour or stripes and do not eat them up. If the sharks eat up the cleaner fishes, then they will not have anyone to clean them up and will get infected. The sharks even let these cleaner fishes to enter their mouths and eat meat particles stuck in their teeth. The cleaner fishes get food and the sharks get a spa service. However, this kind of reciprocal altruism is not seen in the open sea. In the open sea, the probability of a repeated encounter is very little, which is not the case within a closed marine society. The sharks play a repeated game and do not want to eat up the cleaner fishes in order to be able to return to the same spa again and again. Reciprocal altruism is seen only when there are high chances of the repeated encounter.

Live and Let Live

Christmas Truce of 1914 is a piece of well-documented history. A war zone indeed is not a place to look for human cooperation. But that is exactly what was seen during the First World War—a time when trench warfare was the primary means of protecting and gaining ground. The German and English soldiers celebrated Christmas together in 1914, which in military terms could have been deemed as fraternization with the enemy—a serious military offence that calls for court martial. The respective high commands overlooked it because of the involvement of a large number of soldiers including officers. Both the British and German press tried to hush up the incidence till *New York Times* reported it on 31 December 1914. General Sir Horace Lockwood Smith-Dorrien, the Commander of British II Corps, issued warnings against such 'friendly communication' in future. But unofficial ceasefire, which became known as 'live and let live', became norms along the Western Front. It started with an understanding of not launching attacks during mealtime and washing time, and eventually got extended to longer periods. Even when the soldiers launched attacks, they did not mean to kill the opposition. The soldiers understood that they could live if they did not kill. This reciprocal adoption of 'live and let live' was the only way they could have any control over their own lives, and it was possible because the soldiers were stationed in the same trenches for long periods. Understanding the 'live and let live' strategy adopted by the soldiers, the generals ordered rapid shuffling of troops along the front. The cooperation of 'live and let live' between enemy soldiers was possible because the game was a repeated game. With the new system, the soldiers were unable to develop the reciprocal relationship with the enemies as the game lengths were considerably shortened.

Trigger Strategies and Repeated Games

The idea of a trigger strategy was incorporated in each of the examples of reciprocal altruism narrated in the previous section. There may be different variants of trigger strategies, but in general they are constructed on the following three fundamental principles:

1. Begin with cooperation.
2. Cooperate as long as the other(s) do so.

3. Upon observation of defection by any other player, punish them by reverting to non-cooperative play for a pre-specified period.

Fundamentally, it is a cooperative strategy. To ensure cooperation from other players, it uses threats of future punishment that can be executed in a repeated game. In order to see how trigger strategies work, let us walk through an example of a trigger strategy.

Grim Trigger Strategy

The grim trigger strategy (GTS) is an extreme form of trigger strategy, which is similar to the strategy used by the 'grudger' birds. It starts with cooperation and cooperates until the other player defects. Upon observing a defection, it reverts to the non-cooperative play and never returns to cooperation. It is a trigger strategy with longest memory and is extremely unforgiving. If the trigger is drawn, it ensures mutual destruction and that is why often compared to mutually assured destruction (MAD). It hurts the punisher too harshly and may seem to be non-credible. In order to understand how the GTS works, let us consider a pricing game between two firms producing products that are perfect substitutes. For simplicity, suppose that there are only two feasible prices for the products—₹105, which gives the bare minimum 5 per cent profit to the firm and ₹108, which gives a slightly better 8 per cent profit to the firm. The market is very competitive as the consumers are extremely price sensitive. Whoever charges the lower price gets the entire market. The market is matured and is not growing at all. Suppose that the market demand in any quarter is 10 million units if at least one of the firms sells at ₹105. If both firms price it at ₹108, demand reduces to 8 million. If the firms charge the same price, the market demand gets equally shared. Prices are determined at the beginning of a quarter and the firms are committed not to change the price during the quarter. The payoff matrix for the one-shot game (if there was just one quarter to bother about) is given in Figure 6.1. Payoffs are profits in millions of rupees.

Figure 6.1: Payoff Matrix for the One-shot Game

Firm B

		₹*105*	₹*108*
	₹*105*	25, 25	50, 0
Firm A			
	₹*108*	0, 50	32, 32

From Figure 6.1, we can see that pricing their respective products at ₹105 is the strictly dominant strategy for both the firms. The strategy of both the firms charging ₹105 is the Nash equilibrium of the game, but both charging ₹108 is a better outcome for the industry as well as for both firms vis-à-vis the Nash equilibrium outcome. Had the firms colluded to price their respective products at ₹108, they would have earned ₹32 million each instead of ₹25 million. But they cannot collude in the one-shot game in the absence of trust. If Firm A posts its price at ₹108, the best response of Firm B is to price its product at ₹105 and earn ₹50 million. The same is true for Firm A if Firm B posts its price at ₹108. Anticipating that the competitor will defect from the price collusion, the firms will not be able to charge the collusive price, that is, ₹108, and will end up charging the competitive price, which is ₹105. The game is a prisoner's dilemma.

If the game is not a one-shot game but a repeated one, which is more likely for firms competing in a matured market, the firms could have colluded at the ₹108 price using a GTS. Here, the firms need a communication between them. Suppose Firm B initiates a proposal of price collusion using a communication like the following at the beginning of a quarter:

We will price our product at ₹108 this quarter. If Firm A also price their product at ₹108, we will price our product at ₹108 next quarter too and continue to price our product at ₹108 in all subsequent quarters till Firm A reduce it below ₹108. If in any quarter Firm A reduces the price to below ₹108, we will revert to the competitive price of ₹105 from the following quarter.

This communication is using a GTS. Let us see if it works. To understand the working of GTS let us try to analyse and see what Firm A will do.

Firm A can very well ignore the communication and continue to post its product's price at ₹105. In that case, Firm A earns ₹50 million in that quarter and ₹25 million in all subsequent quarters. Firm A can agree to collude on price, but in that case it should also pose a similar counter-threat to Firm B to ensure that Firm B does not defect.

Given Firm B's communication, if Firm B adheres to the proposed GTS, Firm A earns ₹32 million in all quarters by agreeing to the collusion. Consider the first two periods. Firm A earns ₹64 million in the first two quarters vis-à-vis ₹75 million that it would have earned had it ignored Firm B's communication. Clearly, being rational, Firm A will not agree to Firm B's proposal of collusive pricing at ₹108, if it is extremely myopic and sees the payoff of just two quarters. Even if it calculates the payoffs of the first three quarters, it will not collude. We can see that using the GTS it is not possible to collude in this game if at least one of the firms makes decisions on the basis of payoffs in very near future. In order to be able to collude, and in order to sustain the collusion, the firms need to see the game as a repeated game. That is possible if the firms develop a long-term market relationship. In this game, given Firm B's communication, Firm A's payoff from collusion exceeds its payoff from competition if the game is repeated for at least four quarters. If Firm A prices its product at ₹105, it will earn ₹50 million in the first quarter and ₹25 million in the next three quarters. Instead, had it priced its product at ₹108, it would have earned ₹32 million in each of the four quarters. So, if Firm A thinks that the game will be repeated for at least four quarters, it will not defect in the first quarter. Will Firm A defect in the second quarter? By doing so Firm A will earn ₹32 million in the first quarter, ₹50 million in second quarter and will earn ₹25 million in each of the next two quarters, taking their total in four quarters to ₹132 million, which is more than the ₹128 million that they would have earned had they chosen the collusive price in all four quarters. So, if Firm B knows that Firm A perceives the game as a four quarter repeated game, it should not trust that Firm A will not defect. Knowing that it cannot trust Firm A, Firm B will not propose the collusion using the GTS. Since the game is symmetric, even Firm B will have incentive to cheat if it perceives the game to be a repeated game with four or less number of periods.

Does it mean that they will be able to collude if they perceive the game as a repeated game with more than four repetitions? Suppose that both Firms A and B know that the market will cease to exist after five quarters. Will they be able to collude? If they collude and hold on with the collusion for the five quarters, each firm will earn ₹160 million. Can't one of them outsmarts the other by defecting in between? A simple math shows that both firms will have incentive to cheat in the third quarter. If Firm B plays its GTS and keeps pricing the product at ₹108, Firm A can earn a total of ₹164 million by defecting in the third quarter. The same is true for Firm B if Firm A adheres to the GTS. So, both will try to outsmart the other by defecting in the second period. Check the math. If Firm A defects in third quarter, being a sitting duck, Firm B will earn ₹32 million in the first two quarters, nothing in the third quarter and ₹25 million in the fourth and fifth quarters, which takes its total over five quarters to ₹114 million. Of course they won't be a sitting duck. Anticipating that Firm A will defect in the third quarter, Firm B should price its product at ₹105 in the third quarter itself. That way it earns ₹25 million in the third quarter too and its total payoff increases to ₹139 million. It is even better for Firm B if they defect in the second quarter itself. If they defect in the second quarter they earn ₹32 million in quarter one, ₹50 million in quarter two and ₹25 million in the last three quarters, taking its total to ₹157 million. This way Firm B can outsmart Firm A. But Firm A too will not be a sitting duck. It too will anticipate that Firm B will defect in second quarter and it itself will price its product at ₹105 in second quarter itself. Just do the math and you will see that Firm A does even better if it cheats in the first quarter itself. Firm B will be able to anticipate that and it too will price its product at ₹105 in the first quarter itself. So, the GST does not help in colluding and sustaining the collusion even if the game is repeated for five quarters. In fact, it is not possible to collude and sustain collusion if the players know when the game will stop repeating. If the game length is fixed and known to the players involved, it is not possible to sustain collusion using the GTS.

In order to be able to collude and to sustain collusion, the players must think that the game is going to be repeated infinitely. If the players know when exactly the game will stop repeating, then they can look forward and reason backward. When the players are able to apply this logic of backward induction, it is not possible to collude and sustain collusion even using the GTS. But in reality, people cannot apply backward induction in repeated games since they don't know when exactly the game will stop repeating. If the players don't know the game length, they do their calculations looking forward. However, if the players don't know when the game is going to end and they perceive the game to continue infinitely, then it seems that there is nothing to calculate. On the face, it seems that the players' payoffs sum up to infinity if the game is repeated for infinite periods. Are we missing out something? Yes, we need to consider time value of money.

By agreeing to collude, vis-à-vis competitive pricing, Firm A sacrifices ₹18 million in the first quarter to earn ₹7 million more in all subsequent quarters. Firm A won't collude if the PDV of ₹7 million in all subsequent periods is less than ₹18 million. Before we proceed, we need to understand the concept of PDV. Suppose ₹1 invested today gives a quarterly return at the rate r. So, ₹1 invested today will become ₹$(1 + r)$ after one quarter, ₹$(1 + r)^2$ after two quarters and so on. This in turn means getting ₹1 after one quarter is the same as getting ₹$[1/(1 + r)]$ today. That ₹$[1/(1 + r)]$ will become ₹1 after one quarter. Hence this ₹$[1/(1 + r)]$ is the PDV of ₹1 earned one quarter later. Similarly, the PDV of ₹1 earned after two quarters is ₹$[1/(1 + r)^2]$ and so on. Let us denote $[1/(1 + r)]$ as δ, which is called the discounting factor. Now, the PDV of a cash flow of ₹1 earned each quarter infinitely is the following sum in rupees:

$$1 + [1/(1 + r)] + [1/(1 + r)^2] + \ldots = 1 + \delta + + \delta^2 + \ldots = 1/(1 - \delta)$$

Note that $\delta < 1$ for all values of $r > 0$, and for all values of $\delta < 1$, the PDV is a finite number. So, even if the game is repeated infinitely, the PDV of the stream of payoffs earned by the players will be finite.

Given Firm B's GTS, if Firm A chooses to collude and price its product at ₹108 in the first quarter and each subsequent quarter till Firm B defects, then Firm A earns ₹32 million in each quarter. If they defect in the first quarter, they earn ₹50 million in that quarter and ₹25 million in all subsequent quarters. By defecting, Firm A earns ₹18 million more in the first quarter and earns ₹7 million less in all subsequent quarters. If the PDV of the cash flow of ₹7 million per quarter from the second quarter is less than ₹18 million, Firm A will defect in the first quarter. In fact, the calculation remains the same for defection in any quarter, as they will earn ₹18 million more in the quarter they defect and will earn ₹7 million less in all subsequent quarters. The PDV of the cash flow of ₹7 million per quarter from the second quarter is ₹$[7\delta/(1 - \delta)]$ million. If $18 > [7\delta/(1 - \delta)]$, that is, $\delta < 18/25$, that is, $[1/(1 + r)] < 18/25$ or $r > 7/18$, then Firm A will defect in the first quarter. $r > 7/18$ means that the quarterly rate of return on investment is more than 38.89 per cent. At this rate of return, ₹18 million invested in the first quarter will generate a cash flow of more than ₹7 million per quarter for all subsequent quarters, and, in that case, Firm A has no reason to collude with Firm B. Since the game is symmetric here, there won't be any reason for Firm B to collude if their quarterly rate of return on investment is more than 38.89 per cent. If the discounting factor for both the firms are more than 18/25, that is, if the quarterly rate of return on investment is less than 38.89 per cent for both of them, then it is possible for them to collude and sustain collusion using GTS, provided they do not know when the game will stop repeating.

For any game, given the payoffs, we can find a critical minimum value of the discounting factor, and hence a critical maximum periodic rate of return on investment for which it is possible for the players to collude and sustain collusion using a GTS.

There are other examples of trigger strategies as well. If a GTS is an extreme form of the trigger strategy with the longest memory, tit-for-tat (TfT) is the other extreme with the shortest memory. If a player adopts TfT, s/he will just do what the rival did in the immediately preceding period. In our pricing game example, if Firm B adopted TfT, its communication at the beginning of a quarter would have been:

> *We will price our product at ₹108 this quarter. From next quarter, in all subsequent quarters, we will price it at ₹108 if Firm A priced their product at ₹108 in the previous period, and we will price it at ₹105 if Firm A priced their product at ₹105 in the previous period.*

TfT is the most forgiving strategy. In the case of the GTS, if the trigger is drawn, the game becomes non-cooperative forever. But in the case of TfT, the game may return to cooperation if the defector repents and corrects himself or herself.

Any player, confronted with a trigger strategy, will weigh the gains from defection vis-à-vis the loss due to punishment. Since the gain is immediate and the loss is in future, the players will compare the PDV of future losses against the gain and decide to defect if the present gain is larger than the PDV of future losses. To compare the deterrent effects of threats of punishment in different trigger strategies, we need to do the math in the context of the game. The basic process of doing that math is the same as done in the case of the GTS. Interested readers may refer to any post-graduate level textbook in game theory, for example, Osborne (2004) or Fudenberg and Tirole (1993). In the next section, we will discuss some business applications of trigger strategies in repeated games.

Trigger Strategies: Lessons from Theory and History

Form our discussion in the previous section, we understood that in a repeated play, trigger strategies succeed in posing credible threat and developing trust amongst the players under certain conditions. The conditions as per our theory developed in the previous section are:

- The players must understand that the game is going to be repeated for a very long time that is, they cannot see the end of the game.
- For the players, the immediate gain from defecting should not outweigh the loss due to punishment.

Here, we will intuitively discuss three more conditions that are required for the stability of collusion.

Reaction Lag of Players

Collusion gains stability if the reaction lag of the players is short. If a player knows that his or her defection will be noticed immediately and the rival(s) will react immediately, he or she is unlikely to defect. For most product categories, we observe a very high degree of price alignment across the electronic retailers. If a retailer reduces its price, it becomes visible to all others immediately and they react. This makes the gain from defection very small. On the other hand, if the players know that rivals won't be able to react immediately, they tend to defect. A longer reaction time from rivals increases the gain from defection. For example, project contractors can hardly collude unless they bid for similar contracts day in, day out. Even if the players know that they will bid for similar contracts in future, they cannot trust each other that they won't undercut price. The immediate gain from undercutting price is huge and, hence, defection is likely.

Number of Players

The number of players colluding should be small for the collusion to be stable. Suppose that a large number of firms in a competitive market agreed to collude on a price that is more than the competitive price. Since the market is competitive, that is, the buyers are price sensitive, any firm that undercuts price will get the entire market as against the fraction of market they were getting by selling at the collusive price. So, by defecting they increase sells and thus revenue. If revenue did not increase they wouldn't have defected. The increase in sales and the consequential increase in revenue from defection are larger if the number of firms increases. So, the gain from defection is more if there are a large number of firms. Knowing that the incentive to defect is high, the firms will not be able to trust each other and, hence, will not be able to collude.

Capacity of Firms

If the total capacity of firms is considerably less than the market demand, it is easier for firms to sustain collusion. Particularly, if at the collusive price, the firms are operating at near full capacity utilization, then there is no incentive to cut price. By cutting price, a firm is able to increase demand for its product, but if they don't have the capacity to produce, then there is no gain from defection.

Lessons from OPEC

Organization of the Petroleum Exporting Countries (OPEC) provides a good example of the trigger strategy in practice.

Case Study 6.1: Crude Pricing by OPEC

OPEC was formed in 1960 as a cartel between five crude oil exporting countries—Iran, Iraq, Kuwait, Saudi Arabia and Venezuela. At present, there are 12 member countries of OPEC—Algeria, Angola, Ecuador, Iran, Iraq, Kuwait, Libya, Nigeria, Qatar, Saudi Arabia, the United Arab Emirates and Venezuela.

(Case Study contd.)

(Case Study contd.)

The stated mandate of OPEC is to 'coordinate and unify the petroleum policies' of its member countries to 'ensure the stabilization of oil markets' in order to 'secure an efficient, economic and regular supply' to consumers, 'a steady income to producers' and a 'fair return on capital for those investing in the petroleum industry'.

As of 2013, 81 per cent of world's proven oil reserves were located in OPEC member countries, amounting to 1,207 billion barrels. However, OPEC's market share is about 40 per cent. Clearly, OPEC is in no hurry to drill its reserves. The oil quality is not the same across the output of different OPEC countries. Venezuela and Iran produce heavy crude vis-à-vis the light crude of Saudi Arabia, Iraq and Nigeria. OPEC Reference Basket of Crudes (ORBC) price is a weighted average price of crude produced by different member countries. Having 40 per cent market share, ORBC price is an important benchmark in the oil industry. In order to have control over global crude price, OPEC must have perfect control over ORBC, and in order to do that OPEC sets weekly crude output quota for its member countries.

If a member country defects from its given quota, then that country is first warned, and if they continue to defect, the country's membership of OPEC is suspended for a period. Later, evaluating that country's production pattern during the period of suspension, OPEC decides whether to renew the country's membership or to terminate it. Ecuador was suspended from December 1992 until October 2007. Indonesia is suspended since January 2009. The membership of Gabon was terminated in 1995.

Source: Author.

OPEC controls the quantity of crude that OPEC member countries supply to the market by setting weekly quotas for each country. They succeed in meeting two objectives. First, they are able to control the depletion of their reserves. Second, they are able to keep prices high by supplying less. This way the OPEC countries can have a sustained cash flow from oil revenue over longer time horizon. But in the short run, individually each member country may have incentive to defect from the cartel. Many of these countries are entirely dependent on their oil revenue. If a member country runs into deep fiscal deficit, the country will naturally want to meet the deficit by increasing its oil revenue. If that country jacks up its output, there will be an impact on price. The price will fall, but not drastically if other countries abide by the quotas given to them. The defecting country will be able to increase revenue from the increase in output, despite the fall in price. But the quota abiding countries will suffer losses in revenue due to price fall. If the other countries too start overproducing to recover the loss of revenues, price will plummet further and reserves will get depleted faster.

The threat of suspension or termination of membership is a threat against defection. But how credible is that threat? For a country in fiscal crisis, cash in hand is worth much more than the future cash flow. Such a country may realize that the gain from defection is more than the cost due to the punishment. Is termination of membership at all a punishment?

The countries who had been repeated offenders are all located far from the Persian Gulf and Middle East region. Ecuador in South America, Indonesia in Southeast Asia, Gabon in West Africa and so on. There isn't even a single example of a Middle Eastern country defecting from their quota during the period 1982–2014. Why? The answer could be hidden in regional cooperation in oil transportation and other economic fronts. Many of the Gulf countries including Saudi Arabia, Iraq, Kuwait and Qatar share (or shared in some cases) cross-border pipelines. Everyday about 17 million barrels of crude pass through the Strait of Hormuz, which is strategically the most important oil chokepoint on the planet. Everyday about 14 wide tankers pass through a 3.2 km wide shipping lane through the territorial waters of Iran and Oman. Of 17 million barrels passing daily through this chokepoint, Saudi Arabia alone sends approximately 6.3 million barrels. This won't be possible without the cooperation from Iran. For the countries located in the Persian Gulf, the cost of defection is much higher than the ones located in South America or West Africa or Southeast Asia.

Another factor plays an important role. Most of the defectors had very little oil reserves. Venezuela might be located in South America, far from the Persian Gulf, but they have the largest reserve. Venezuela alone has 28 per cent of the world's proven oil reserve. With that a big reserve, Venezuela's interests are aligned with that of OPEC. So far Venezuela didn't defect repeatedly and had never been suspended. But with a rapidly changing economic condition in the country, at the time when this book is being written, one won't be surprised if Venezuela defects. Nigeria may not have a huge reserve like Venezuela or Saudi Arabia, but they are the largest producer of crude in Africa and they have one of the largest reserves of light sweet crude—the more expensive and sought after variety that is used in the production of gasoline. The countries with larger reserves perceive the oil pricing game as an infinitely repeated one. Hence, they care for the future and don't want to defect, as the membership of OPEC is more valuable to them. On the other hand, for countries with smaller reserves, the immediate gain offsets the future losses due to punishment.

OPEC's success is manifested by the steady increase in crude price from US$30 per barrel in 1998 to US$140 per barrel in 2008. During the global recession, the oil price came down to US$46 per barrel in 2009 but recovered to US$120 per barrel in 2011 and was stable around that price till 2013. Since April 2014, oil price reduced drastically and went below US$50 by the end of 2014 owing to overproduction by OPEC and other oil producing countries. Indeed, the global supply of crude increased due to shale oil boom in the USA. Using fracking technology, the USA had been drilling shale oil in North Dakota and Texas since 2008. US production increased steadily from 8 million barrels per day in 2010 to close to 12 million barrels in 2014. But crude price wasn't falling before April 2014. It is true that Chinese demand reduced in 2014, but global demand increased by 2 million barrels per day as compared to that in 2013. The fall in price can only be attributed to OPEC's failure to control supply. Before April 2014, OPEC was controlling price by controlling supply. Question is why OPEC failed to control supply during the latter part of 2014.

On 27 November 2014, OPEC went for a vote on whether to cut production. The decision based on the outcome of the vote was a clear 'no'. OPEC decided to maintain production at 30 million barrels per day. Normal reaction to falling prices due to increased supply would have been production cut. In the past, if a member country defected, being the largest member country in terms of capacity, Saudi Arabia used to cut down the output and, thus, OPEC used to arrest the fall of price. A smaller member country could be threatened of suspension or expulsion and for Middle Eastern countries it always worked. For member countries located far away from the Persian Gulf, such threats failed to induce cooperation at times. But those were all small producers who cannot impact the market drastically. But the impact of the USA, with a capacity of 40 per cent of that of OPEC, is much stronger. If OPEC cut down production, they will lose market share. It has become a 'war of attrition' where both are playing the holding game. As discussed in Chapter 4, in a 'war of attrition', both players lose money. But they are ready to bear the cost of war, hoping that the rival will throw in towel and give up. The game continues till one of the players gives up. It is a repeated game with a difference. The game repeats depending on the outcome of the stage game. If in a stage game, one of the players gives up, the game is over. But if both hold on, the game continues.

Leveraging Repeated Play to Out-think Customers

Cartelization, as done by OPEC, is an obvious way of leveraging on repeated play. Trigger strategies don't work unless the threat is credible. We understood that from the experience of OPEC. Successful sellers' cartels can keep prices high. Forming explicit cartels like OPEC is impossible for private firms as anti-competitive laws prevent the formation of such cartels in most countries. But cartelization is difficult to prove in the court of law if the firms form implicit collusion without having a written down contract or mandate.

Price Fixing by Implicit Collusion

Seller's collusions fix price and retain high price in pretty much the same way as OPEC, but they work without any written down constitution. Typically, firms producing commodities that are inputs of production for other businesses are often alleged to have engaged in price fixing by means of forming implicit collusion. Case Study 6.2 sums up various instances when DuPont, the manufacturer of various primary chemicals, has been alleged to have engaged in price fixing.

Case Study 6.2: Price Fixing by DuPont and Its Competitors

DuPont Co., Kronos and Cristal USA Inc. are all producers of titanium dioxide, which is an ingredient of paint and, hence, is an input for paint manufacturers. In 2010, in the Maryland Federal Court, a group of titanium dioxide buyers led by Haley Paint Co., Isaac Industries Inc. and East Coast Colorants sued DuPont Co., Kronos and Cristal USA Inc. The plaintiffs alleged that the titanium dioxide producers, who were supposed to be competitors, held secret meetings to exchange sensitive commercial information on their sales, supply, production and pricing. The plaintiffs complained that such meetings were held with the goal of artificially driving up the price of titanium dioxide and to allocate shares of the US market. The alleged co-conspirers all reached settlement in 2013 with DuPont forking over US$75 million. Kronos and Cristal USA Inc. had to cough up US$35 million and US$50 million, respectively, to settle the claims in September 2013.

Even before the case was settled in Maryland, in March 2013 the same group of chemical firms were again sued in a California court by a group of plaintiffs who the court termed as 'indirect buyers'. DuPont Co., Kronos, Huntsman Corporation and Cristal USA Inc. were again sued in a putative class action wherein the plaintiffs alleged that the chemical manufacturers conspired to fix the price of titanium dioxide. It was alleged that

(Case Study contd.)

(Case Study contd.)

> the group of firms had been manipulating the price of titanium dioxide since early 2002.
>
> Source: Haley Paint Co. v. DuPont Co. et al.; Case No. 1:10-cv-00318 in the US Federal Court of Maryland.
> Los Gatos Mercantile Inc. et al. v. DuPont Co. et al.; Case No. 3:13-cv-01180 in the US District Court for Northern District of California.

A chemical like titanium dioxide is essentially a commodity and the producers cannot differentiate their products much. The buyers are businesses who would want to reduce their costs and, hence, they are expected to be extremely price sensitive. As a result, the market will be intensely competitive. In such a scenario, the sellers are forced to play a prisoner's dilemma game wherein they are forced to keep prices low. If a firm reduces price, that firm takes away the market from its competitors, and knowing that the competitors also have to keep prices low. The scenario is fundamentally identical to the game scenario between Firms A and B discussed in the 'Trigger Strategies and Repeated Games' section. Such scenarios are breeding grounds for price collusion. Without collusion between the sellers, it becomes a buyer's market. Unless they collude, the bargaining power is lopsided in favour of the buyers. We shall return to discuss bargaining power in Chapter 8. Collusion helps the sellers to regain the bargaining power. So, it is a survival strategy for the sellers.

As indicated in the Case Study 6.2., DuPont, Kronos, Cristal USA Inc. and other producers of titanium dioxide possibly had communication amongst themselves for framing some sort of a trigger strategy to induce cooperation. The communication might be very similar to that between Firm A and Firm B, discussed in the 'Trigger Strategies and Repeated Games' section.

Most Favoured Customer Clause

The most favoured customer (MFC) clause is generally used in B2B procurement contracts. The contract specifies price, quantity, timeframe, etc., along with the MFC clause. The MFC clause says that the seller treats the customers, with whom it has signed the MFC clause, as its most favoured customers. The MFC clause makes the following promise to customers:

> *If I ever offer a lower price to any other customer, I will offer it to all of my customers as well with retroactive effect.*

The contract is generally valid for a limited period.

Suppose that a firm signs a contract with Buyer 1 to deliver 1,000 units of a product per month at a price of ₹100 per unit, for a period of six months. If the contract has an MFC clause, it would mean that the seller will return money retroactively to the buyer if it sells the same product at a price lower than ₹100 to any other customer during that six month period. Suppose that, after three months, the seller signs another contract with Buyer 2 to sell the same product at a price ₹80 per unit. In such a circumstance, not only the seller will have to deliver the product at ₹80 per unit to Buyer 1 for the remaining three months of the contract, but will also have to give back Buyer 1 a sum of ₹60,000, calculated as the difference in prices of ₹20 per unit for all the 3,000 units delivered in the first three months.

On the face, the MFC clause looks like a buyer-friendly clause. But actually, the buyers are being held as hostages here in order to gain credibility to the competitors. A firm signing contracts with its customers with the MFC clause indirectly makes promise to its competitors that they will keep prices high. In such scenarios where firms can easily poach competitor's customers by offering slightly lower price, the MFC clause restricts the seller from doing so. With the MFC clause in place, the gain from undercutting price gets reduced as the seller is required to pay back money to its existing customers. If the quantum of sale to the existing customers, who have the MFC clause in their contracts, is sufficiently large, the gain from undercutting price may be completely offset by compensation made to the existing customers due to a breach of the MFC clause. In such a scenario, the seller will not reduce price at all. The MFC clause works in the same manner as the burning the bridge strategy discussed in the

'Making Commitment by Elimination of Options' section of the last chapter. It is a strategic move to commit not to reduce price. Even a monopolist may use the MFC clause in contracts with customers to make a commitment of not reducing price. In a tightly competitive scenario, the MFC clause prevents the firms from getting into a price war. The firms realize that in a repeated game, it does not make sense for them to undercut price. If competitors sign contracts with the MFC clause with their respective customers, it becomes easier for them to retain a high collusive price. In the 'Trigger Strategies and Repeated Games' section, we understood that collusion becomes sustainable only if the rate of return on investment is sufficiently low. If the rate of return on investment is not critically low, and the gain from undercutting price offsets the PDV of loss in cash flow resulting from the breakdown of the collusion, the MFC clause can act as a saviour for the collusion. In the presence of the MFC clauses, the gain from undercutting becomes much smaller.

A narrative of a series of appeals and counter appeals in an antitrust case regarding the use of the MFC clause by DuPont and Ethyl Corporation is given in Coopetition (1997) by Brandenburger and Nalebuff. Ethyl Corporation and DuPont supplied lead-based antiknock additives to gasoline producers. They were one of the first firms to use the MFC clause in the competitive scenario. Ethyl Corporation was the pure monopoly before DuPont began producing the antiknock additives, and Ethyl Corporation signed contracts with their buyers even then. The product being an industrial chemical was undifferentiated. If they competed with each other, they would have sold at a very thin margin. The game was indeed a repeated game and getting into a tacit collusion using a trigger strategy was the obvious choice. Being a multi-product firm, the rate of return on investment for DuPont was high. Ethyl Corporation required a promise from DuPont that they will not undercut price. Both companies agreed to sign the MFC clause with their respective customers. In 1979, the FTC of the USA ruled that the MFC clause is anti-competitive. But in 1984, New York Federal Court of Appeals overturned the FTC ruling stating that Ethyl Corporation had MFC clauses with their customer even when they had no competition. It was also argued that buyers gain from the MFC clause as they are ensured that their competitors will not get a better price than them.

Indeed, the MFC clause tilts the bargaining power in favour of the seller. But it is difficult to say that the clause helps in retaining high price. If the quantum of sales from the existing contracts is not very large, the MFC clause as such does not prevent the sellers from reducing price. Moreover, if a firm knows that its competitor signed contracts with their customers with the MFC clauses, it gets encouraged to cut price knowing that it will be difficult for the competitor to react with a similar price cut. The MFC clause is useful in retaining high price, only if all the firms sign contracts with their respective customers using the clause.

Meet the Competition Clause

Meet the competition clause (MCC) requires the sellers to sell their products at the best price offered by any other seller. The MCC is typically used in mass markets, particularly in retail. It makes the following promise to the customers:

If you get a lower price from my competitors, I will match it with retroactive effect.

The offer is generally valid for matching offers with a limited number of competitors and over a limited period, excluding discount sales.

If the MFC clause acts as a promise to the competitors not to reduce price, the MCC acts as a threat to competitors that they cannot get away by cutting price. If a retailer offers the MCC to its customers, its competitors cannot gain by undercutting price. Knowing that they won't gain from undercutting price, the competitors restrain from initiating a price war. If a group of retailers want to collide on a price higher than the competitive price, it becomes easier for them to sustain the collusive price when each of them offers the MCC to their respective buyers. That is exactly what some retailers like Toys"R"Us and Kmart do.

Case Study 6.3: Price Match Guarantee by Toys"R"Us

Toys"R"Us, one of the largest retailers of toys, offers price-match guarantee that reads as follows: *Simply come in store and show us the same item in a competitor's printed ad, selected online retailer's website or our website.* The company extends the offer to match prices from selected online competitor websites, including Walmart.com, Target.com, BestBuy.com, Sears.com, Kmart.com, buybuyBaby.com, Meijer.com, FredMeyer.com, diapers.com, BabyDepot.com, LEGO.com and Amazon.com. But they lay down the following conditions too:

- Price-match guarantee is valid for in-store purchases only.
- Price-match is given at the time of purchase or within seven days of the purchase date with a valid receipt.
- The original, complete competitor ad, with valid dates, must be presented at the time of purchase. Advertisements presented via smartphones must be displayed on competitor's website or app.
- Online prices must be verifiable via competitors' websites.
- Prices are matched after deducting any coupon savings and other offers from our price.
- Price-match guarantee is not valid for category-wide or storewide discounts, buy one get one offers, online-only prices, gift with purchase offers, doorbuster items, coupons, online pricing that is limited to one day or less (e.g., one day deal, six-hour sale, evening sale, etc.), online pricing from a third party selling products via a competitor's site, clearance, closeout, damaged product, used, refurbished, open packages or liquidation sales.

Source: www.toysrus.com

The price-match guarantee offered by Toys"R"Us, as given in Case Study 6.3 is essentially an MCC.

In this chapter, we discussed how cooperation helps players even in a competitive scenario, if the games are not played once, but repeatedly. Cooperation requires trust among the players. It is easier to trust another player when you know that you have a way to penalize him or her if he or she defects. Credible threats and promises help in sustaining cooperation between players who are self-interested and individually rational.

7

Business Poker: Playing Games with Limited Information

Poker is a game of incomplete information. For example, if you are a player in a three-card flash between two players, you only know that your rival doesn't have those three cards that you have. But she/he may have any three cards from the remaining 49 cards. That makes poker a game of chance, but there still remains scope of strategic thinking. In a game of poker you cannot out-think your opponent unless you keep your opponent guessing. If you play multiple rounds with the same set of players and you bid only if you get high cards, your opponent(s) will notice that. So, after a few rounds they will always fold if you bid or raise. Instead, if you sometimes bid or raise even with lower cards and lose, your opponents will be left guessing. In a later round, if you get high cards and raise on a big pot, your rivals might be tempted to call (challenge) if they know that you have a tendency to raise even with lower cards. Bluffing is a part of poker and when to bluff is a strategic decision. In fact, game theorists and computer scientists have developed algorithms of poker that work near perfectly. In January 2015, *Nature* magazine reported that computer scientist Michael Bowling and his colleagues at

University of Alberta at Edmonton, Canada, along with Finnish software developer Oskari Tammelin, developed a computer program that plays the two-player "Texas Hold'em" version of poker perfectly. The program even perfected strategic bluffing.

This chapter is not about poker. There are many business and day-to-day life scenarios that are games of incomplete information. Like poker there is a stochastic element in these games, but nevertheless it is possible to be strategic and to out-think rivals. In this book, we have so far discussed sequential move games, simultaneous move games and repeated games. In each form, we have dealt with situations where nothing is left to chance. Here, in this chapter, we will see how an element of chance or uncertainty affects each of these forms of games.

Keep Your Opponent Guessing

To begin with let us revisit one-shot simultaneous move games discussed in Chapters 3 and 4. We discussed games where players have dominant strategies as well as games wherein the players don't have any dominant strategy. In all the classes of simultaneous move games discussed in Chapters 3 and 4, we found Nash equilibrium or equilibria. However, in some simultaneous move games it might seem that there does not exist any Nash equilibrium. To understand it, let us use a simple game.

Heads up or Tails up

Suppose two players A and B are playing this game. Each of them has one ₹1 coin. The coins look identical—one side marked as 'head' and the opposite side as 'tail'. All they need to do is to show their respective coins with either 'head' side up or the 'tail' side up. They don't toss the coins but keep their respective coins in their fists, with either of the faces up. They show their coins by opening their fists simultaneously. If the faces of the coins match, that is, if both of them show 'head' or both show 'tail', then A

Figure 7.1

B

		Head	Tail
A	*Head*	1, −1	−1, 1
	Tail	−1, 1	1, −1

wins. If the faces of the coins do not match, that is, if A shows 'head' and B shows 'tail', or if it is the other way round, then B wins. The loser gives his or her coin to the winner. The payoff matrix for the game looks like the one given in Figure 7.1.

In this game, apparently, there is no Nash equilibrium. If A shows 'head', B's best response is 'tail', but if B shows 'tail', A's best response is 'tail'. Again, if A shows 'tail', B's best response is 'head', but if B shows 'head', A's best response is 'head'. The Nash equilibrium criterion of 'best responses to each other' is not satisfied for any combination of actions. How to play this game? It seems that they might just choose the faces randomly.

Suppose the game is played between A and B for a very large number of times. How can the players maximize their payoffs? If A always chooses 'head', B will always choose 'tail'. So A must not always choose 'head', but mix his choices between 'head' and 'tail'. If there is a pattern in A's choice between 'head' and 'tail' and if B can recognize the pattern, then B will know what to do to win. So, A must mix between 'head' and 'tail' randomly.

If, suppose, after some rounds, B observed that A is choosing 'head' and 'tail' randomly, but is choosing 'head' 75 per cent of the times. What should B do? If B always chooses 'tail', then B maximizes her chance of winning. When A chooses 'head' in three out of four times, by choosing 'tail' throughout, B wins three out of four times on an average. So, A is not best off choosing 'head' 75 per cent of the times. Given that A chooses 'head' 75 per cent of the times, B's best response is to choose 'tail' always, but if B always chooses 'tail', A's best response is to always choose 'tail'.

What if A chooses 'head' half of the times and 'tail' at the other half of the times? Now, B will win half of the times if she always chooses 'head'. However, B will win half of the times even if she always chooses 'tail'. So, when A chooses 'head' 50 per cent of the times, B becomes indifferent to choosing either 'head' or 'tail'. However, B must not always choose 'head', nor must he/she always choose 'tail'. If B always chooses a particular side of her coin, A will always win if he chooses the same side. So, B must not always choose the same side. She also must mix between 'head' and 'tail' randomly. Since the game is symmetric, by the same logic as why A should mix 'head' and 'tail' by choosing each face 50 per cent of the times on an average, B should also choose each face 50 per cent of the times. When B chooses 'head' and 'tail' equal number of times, A cannot take advantage. In this game, when played for very large number of times, an equal mix between available actions keeps the opponent guessing. But there should not be any pattern. If a player chooses 'head' and 'tail' alternatingly, or 'head' at a trot for N times followed by 'tail' at a trot for N times, and if the opponent identifies the pattern, then the opponent will always win thereafter. That is why it is important to randomize. Tossing the coin would result in perfect randomization.

Mixed Strategy Nash Equilibrium

Even in this game, which apparently has no Nash equilibrium, we have a strategy. It is called mixed strategy. For all practical purposes, a mixed strategy is the proportion in which you mix your different available strategies. It is useful to use the right mix when you don't have a definite choice. Practically, it is possible to apply mixed strategy when the game is repeated for a large number of times. Using the right mix helps you to maximize your cumulative payoff in such games. The idea is to mix amongst your different strategies and actions, such that the opponent cannot take advantage of you.

Technically, a mixed strategy is a probability distribution over different strategies in a player's available set of strategies. The objective of a player using mixed strategy is to maximize the expected payoff. Suppose in the game of 'heads up or tails up',

A chooses the face of his coin by tossing it. So, even if the game is played for just once, he is choosing 'head' with a 50 per cent chance and 'tail' with 50 per cent chance, or as it is conventional to say mathematically, he chooses 'head' with probability 0.5 and 'tail' with probability 0.5. He is actually applying mixed strategy.

The mixed strategies of different players in a game constitute a 'mixed strategy Nash equilibrium' if and only if their mixed strategies satisfy the criterion of 'best responses to each other'. In the game of 'heads up or tails up', as we observed, both A and B choosing 'head' 50 per cent of the times and 'tail' 50 per cent of the times satisfy the criterion of 'best responses to each other' and, hence, this constitutes a mixed strategy Nash equilibrium. Even if the game was played just once, both players choosing 'head' with probability 0.5 and 'tail' with probability 0.5 would be the mixed strategy Nash equilibrium. Before we understand why it is so, we need to understand the concept of mathematical expectation and expected payoff.

Expected Payoff

Suppose B always chooses 'tail'. If A chooses his action by tossing the coin, over a very large number of rounds of the game, his coin will come up 'head' approximately 50 per cent of the rounds and 'tail' for the rest of the rounds. Whenever the coin turns up 'head', he gets −1 and if it turns up 'tail', he gets 1. If the game is played for a very large number of rounds, and if A decides by tossing his coin, then A gets −1 for approximately half the rounds and 1 for the remaining half. His cumulative payoff over all the rounds will be zero or very close to zero. Before the game starts, A can expect his cumulative payoff to be zero if he decides to choose the face of his coin by tossing it while B chooses 'tail'. This will be true for any random trial that is repeated for a very large number of times. For example, if a dice bearing numbers 1–6 on its six faces is thrown for 60,000 times, each of the numbers will come up approximately for 10,000 times. So, even before the dice is thrown even once, we expect each of the numbers to turn up 10,000 times.

Mathematical expectation can, however, be applied for random events that are tried just once. Suppose the game of 'heads up or tails up' is played for just once and B chose 'tail'. If A tosses his coin, he will either get −1 (in the event of his coin turning up 'head') or 1 (in the event of his coin turning up 'tail'). But, ex ante, that is, before A tossed the coin, he knew that he will get −1 with probability 0.5 and 1 with probability 0.5. His expected payoff in this case is his actual payoff in each of the two possible events multiplied by their respective probabilities, that is, $[0.5 \times 1 + 0.5 \times (−1)]$. He will never get zero though. Either he will get −1 or he will get 1. By applying mathematical expectation, we can say that his expected payoff from tossing the coin is zero.

Mixed Strategy Nash Equilibrium in the One-shot Game

We proposed that each player choosing 'head' with probability 0.5 and 'tail' with probability 0.5 constitutes a mixed strategy Nash equilibrium in the one-shot game as these probability distributions satisfy the criterion of 'best responses to each other'. Now let us check that proposition using our understanding of mathematical expectation.

Given that A chooses 'head' with probability 0.5 and 'tail' with probability 0.5, B's expected payoff from choosing 'tail' is 0 and that from choosing 'head' is also zero. That makes B indifferent to choosing 'head' or 'tail' and prompts her to randomize as well. Randomizing between 'head' and 'tail' will give her an expected payoff that is only a linear combination of the expected payoffs from choosing 'head' and that from choosing 'tail'. But if B randomizes between 'head' and 'tail' with any probability distribution other than assigning equal probabilities on each action, A will be able to take advantage. Suppose B chooses 'head' with probability p and 'tail' with probability $(1 − p)$. A's expected payoff from choosing 'head' will be $[p \times 1 + (1 − p) \times (−1)] = (2p − 1)$. A's expected payoff from choosing 'tail' will be $[p \times (−1) + (1 − p) \times 1] = (1 − 2p)$. For any $p > 0.5$, $(2p − 1) > (1 − 2p)$, and hence

A will choose 'head'. Similarly, for any $p < 0.5$, $(2p - 1) < (1 - 2p)$ and hence A will choose 'tail'. Hence, if B chooses 'head' with probability p and 'tail' with probability $(1 - p)$ where $p \neq 0.5$, A's best response is not to randomize but to choose 'head' or 'tail' depending on the value of p. When A chooses 'head', and B chooses 'head' with probability p and 'tail' with probability $(1 - p)$ where $p > 0.5$, B's expected payoff is $[p \times (-1) + (1 - p) \times 1]$ $= (1 - 2p)$, which is less than zero. When A chooses 'tail', and B chooses 'head' with probability p and 'tail' with probability $(1 - p)$ where $p < 0.5$, B's expected payoff is $[p \times 1 + (1 - p) \times (-1)] =$ $(2p - 1)$, which is less than zero. So, if A randomizes by choosing 'head' with probability 0.5 and 'tail' with probability 0.5, B's best response is also to randomize and choose 'head' with probability 0.5 and 'tail' with probability 0.5. Since the game is symmetric, we can conclude that when B randomizes by choosing 'head' with probability 0.5 and 'tail' with probability 0.5, A's best response is to do the same. Therefore, each player randomizing by tossing their respective coins constitutes the mixed strategy Nash equilibrium for this 'heads up or tails up' game even if the game is played just once. Indeed, as was mentioned before, applying mixed strategy makes practical sense if the game is repeated for a very large number of times. In that case, by randomizing just right, each player maximizes their cumulative payoffs. In the one-shot game, ex post, that is, after they toss their coins, either both coins will turn up 'head' or both will turn up 'tail', or one will turn up 'head' while the other turns up 'tail'. The players will either lose their coin or gain the opponent's coin depending on the outcome. But still tossing the coin to take decision makes scientific sense for them, provided they are 'risk neutral'. Let us understand individual risk behaviour in the next section.

Individual Risk Behaviour

Consider a simple lottery. There are 52 cards in a well-shuffled deck. Out of which, 26 are red (hearts and diamonds) and 26 are black (spades and clubs). If you draw a black card you will get ₹200, but if you draw a red card you will not get anything. No harm in gambling in such lotto if you lose nothing. But normally for any such lottery you need to pay a participation fee. What is the maximum amount that you are willing to pay to participate in the lottery mentioned above? The answer will vary from person to person.

The probability of drawing a black card is 26/52, that is, 0.5 and that of drawing a red card is also 0.5. When a black is drawn you get ₹200 and when a red is drawn you get nothing. This means you will get ₹200 with a probability 0.5 and nothing with probability 0.5. So, your expected payoff from the lottery is ₹100. If you are willing to pay more than ₹100 to take part in this lottery, you are a risk-lover. If you are willing to pay exactly ₹100 to take part in this lottery, you are a risk-neutral, and if you are willing to pay less than ₹100 to take part in this lottery, then you are risk-averse. The amount you are willing to put up to take part in a lottery is called the certainty equivalent of the lottery. For a risk-neutral individual, the certainty equivalence is equal to the expected payoff. For a risk-lover, it is more than the expected payoff and for a risk-averse individual, it is less than the expected payoff.

The certainty equivalence of a risky prospect (a lottery is a prospect) can be also interpreted as the certain amount that will dissuade an individual from the risky prospect. A risk-averse person easily gets dissuaded by the risk. She/he prefers a certain amount over the risky prospect even if the certain amount is less than the expected payoff from the risky prospect. A risk-lover, on the other hand, is attracted by the high return associated with the risk. Therefore, a risk-lover requires more than the expected payoff with certainty to be dissuaded from the risky prospect.

When an individual takes a risk for another individual, the one who takes risk needs to be compensated to cover for the risk. For example, in some professions like fire-fighting or flying or mining, there are life risks of the workers. The workers in these professions get paid more than what they would have been paid had there been no life risks. The extra amount that the workers are paid for taking risk is called the risk premium. Even in case of buyer–seller contracts under uncertainty, if the entire liability of risk due to uncertainty is taken by the seller, the seller is required to be paid a risk premium. Since it is harder to pursue a risk-averse individual to take risk, they demand a higher risk premium as compared to risk-lovers or even risk-neutral ones.

In this chapter, we will mostly assume that our game players are risk-neutral. Risk-neutral individuals become indifferent to a risky prospect or a certain amount if the certain amount is equal to the expected payoff from the risky prospect. In other words, they simply maximize their expected payoffs.

Mixed Strategy in Anti-coordination Games

'Heads up or tails up' is a game without any Nash equilibrium in pure strategy. Pure strategy refers to strategies without any mixing, as opposed to mixed strategy, which is a probability distribution over the set of pure strategies. But even in games with multiple Nash equilibria, the players might mix their pure strategies, and mixed strategy Nash equilibrium may exist. We discussed coordination games as well as anti-coordination games in Chapter 4. Both in coordination games and anti-coordination games there exist multiple Nash equilibria, and in both classes of games, there exist mixed strategy Nash equilibria.

Consider, for example, the BoS game discussed in the 'Coordination Games' section of Chapter 4. The payoff matrix of the game is given in Figure 4.1. In this game there exist two Nash equilibria—(football, football) and (concert, concert). The woman prefers the former equilibrium and the man prefers the later. But since they both are better off in any of the Nash equilibria vis-à-vis the non-Nash equilibrium outcomes, they want to coordinate and reach one of the Nash equilibria. If the game is played for a very large number of times, they can make a pact and coordinate alternatingly on the two Nash equilibria. Hence, in this game, mixed strategy can be used only in the one-shot game where the players randomly choose their actions. That is the scientific approach to make decision, ex ante, but ex post they may end up with one of the Nash equilibria or they might fail to coordinate.

In anti-coordination games, the players don't want to coordinate. Consider the hawk–dove game discussed in the 'Anti-coordination Games' section of Chapter 4. The payoff matrix of the game is given in Figure 4.5. There exist two Nash equilibria—(hawk, dove) and (dove, hawk). A prefers the former equilibrium and B prefers the later. Not only that, but A prefers the (dove, dove) outcome over the (dove, hawk) equilibrium and B prefers the (dove, dove) outcome over the (hawk, dove) equilibrium. The players don't want to coordinate and hence they would randomize. Suppose A chooses to be a hawk with probability p and chooses to be a dove with probability $(1 - p)$. If B chooses to be a hawk, B's expected payoff is $[-5 \times p + 10 \times (1 - p)] = 10 - 15p$. If B chooses to be a dove, B's expected payoff is $[0 \times p + 5 \times (1 - p)] = 5 - 5p$. B will be indifferent if and only if $10 - 15p = 5 - 5p$, that is, $p = 0.5$. By randomizing between being a hawk and being a dove with equal probabilities, A keeps B guessing and hence B cannot take advantage of A. Note that we have found the right mix for A using B's expected payoffs. Similarly, we can find the right mix for B. B will be able to keep A guessing by choosing to be a hawk and choosing to be a dove with equal probabilities. Both players mixing 'hawk' and 'dove' equally constitutes the mixed strategy Nash equilibrium in this game too. Since the players don't want to coordinate, even if the game is played for a very large number of times, the players will do best by mixing 'hawk' and 'dove' equally, but randomly.

In all the examples in this chapter, where we calculated the mixed strategy Nash equilibrium, we found that the players do best by mixing between different actions with equal probabilities. But that is not always the case. Consider another anti-coordination game discussed in Chapter 4. The payoff matrix of the chicken game is given in Figure 4.6. Suppose driver 1 dares and chickens with equal probabilities. Given this mixed strategy of driver 1, driver 2 is clearly better off by chickening out. The expected payoff of driver 2 from daring is $[-5 \times (0.5) + 1 \times (0.5)] = -2$ and that from chickening out is 0. But if driver 2 chickens out with certainty, then driver 1 should rather dare with certainty than mixing between daring and chickening with equal probabilities. Clearly, in the mixed strategy Nash equilibrium, if that exists, the drivers should not mix between their actions with equal probabilities. Let us find out the mixed strategy Nash equilibrium. Suppose driver 1 dares with probability p and chickens with probability $(1 - p)$. The expected payoff of driver 2 from daring is $[-5 \times p + 1 \times (1-p)]$ $= 1 - 6p$ and that from chickening out is 0. So, driver 2 will be indifferent to daring or chickening out if and only if $1 - 6p = 0$ or $p = 1/6$. Even by daring with a probability $1/6$, driver 1 is being able to make driver 2 indifferent to daring or chickening out. In case the game is played for a very large number of times, driver 1 will be able to keep driver 2 guessing by choosing to dare only once in every six times on an average. Similarly, we can check that driver 2 will be able to keep driver 1 guessing by choosing to dare with probability $1/6$. Suppose driver 2 dares with probability q and chickens with probability $(1 - q)$. The expected payoff of driver 1 from daring is $[-5 \times q + 1 \times (1 - q)] = 1 - 6q$ and that from chickening out is 0. So, driver 1 will be indifferent to daring or chickening out if and only if $1 - 6q = 0$ or $q = 1/6$. The game is symmetric and we find a symmetric mixed strategy Nash equilibrium where each driver mixes between daring and chickening with probabilities $1/6$ and $5/6$, respectively. The mixed strategy Nash equilibrium is symmetric, but the two actions are not chosen with equal probabilities. This is because of the huge negative payoff associated with the (dare, dare) outcome, compared to the winner's payoff in the (dare, chicken) or (chicken, dare) Nash equilibria.

In the chicken game, the objective of each driver is to ensure that the other chickens out. For example, if driver 1 dares with a probability higher than $1/6$, then driver 2 will chicken out. But driver 2 can also try to do the same and choose to dare with a probability higher than $1/6$. But that will not constitute a mixed strategy Nash equilibrium as the criterion of 'best responses to each other' will be violated. If driver 1 chooses to dare with probability more than $1/6$, driver 2 will chicken. But if driver 2 chickens out, then driver 1 must dare with certainty. However, if the game is played for a very large number of times and if driver 2 knows that driver 1 dares more often than once in six times, then driver 2 will chicken out. So, by creating a reputation of being hard-headed, it is possible to win in an anti-coordination game. We shall discuss about reputation in the 'Reputation' section further.

Mixed Strategy Applications

There are many real-life business situations where the idea of mixed strategy is useful. In this section, we will discuss a few applications.

Monitoring

Monitoring involves cost to whoever monitors—employer, regulator or even examination invigilator. But in absence of monitoring, the worker will shirk; the regulated firms, organizations or institutions will be involved in malpractices, and examinees might cheat. Let us address the problem with the example of employee monitoring. An employee involved in a production process is supposed to put out a certain quantity of work in a day or hour during his or her regular work hours. If the worker shirks, she/he gets leisure during work hours, which is an added benefit. We can see it as a reduced cost of effort for the worker. So, in absence of monitoring, the worker will shirk. If the job contract involves a penalty against shirking, the workers won't shirk when monitored. For the employer, or the supervisor, it is costly to monitor. If the workers don't shirk, it is better for the employer if they do away with monitoring. But if they are not being monitored the workers will shirk. In order to see how mixed strategy might be helpful, let us construct an example.

Suppose a worker is supposed to put out a certain output in an hour and the cost of effort is equivalent to ₹100 per hour. The wage rate is ₹200 per hour and the value of output produced by the worker in an hour is ₹400. Suppose the cost of monitoring is ₹100. We will consider the worker's payoff as the wage net of

Figure 7.2

Employer

	Monitor	Not monitor
Work	100, 100	100, 200
Shirk	0, −100	200, −300

Worker

effort cost, and the employer's payoff is the value of the output net of wages paid and monitoring cost. When the worker shirks, his or her effort cost is zero and output is zero. But if caught shirking, the worker is not paid wages. The payoff matrix of the game is given in Figure 7.2.

In this game, there does not exist a Nash equilibrium in pure strategies. Let us see if there exists a mixed strategy Nash equilibrium. Let the employer monitor with a probability p. The worker's payoff from working is 100, irrespective of whether the employer monitors. The worker's expected payoff from shirking is $[0 \times p + 200 \times (1 - p)] = (200 - 200p)$. The worker will become indifferent to working or shirking if $(200 - 200p) = 100$, that is, $p = 0.5$. Suppose the worker works with a probability q. The employer's payoff from monitoring is $[100 \times q - 100 \times (1 - q)] = (200q - 100)$ and that from not monitoring is $[200 \times q - 300 \times (1 - q)] = (500q - 300)$. The employer will become indifferent to monitoring or not monitoring if $(200q - 100) = (500q - 300)$, that is, $q = 2/3$. So, in the mixed strategy Nash equilibrium, the worker will work with probability 2/3 and shirk with probability 1/3, whereas the employer will choose to monitor and not to monitor with equal probabilities. A game like this is not one shot. It is a game that is played day in day out. So, the worker's long-term strategies of working in two out of three times as against that of the employer of monitoring once in every two times constitute a mixed strategy Nash equilibrium. Indeed, the employer will want the worker not to shirk at all. To achieve that outcome, the employer needs to monitor more than once in two times. For example, if the employer monitors thrice in every five times, on an average, and does that randomly, the best response of the worker will be not to shirk at all. That does not constitute a Nash equilibrium as the employer's best response is not to monitor if the worker never shirks. But practically, it works if the employer monitors just a little more frequently than once in two times.

War of Attrition Revisited

In Chapter 4, we discussed the case of BSB versus Sky TV war of attrition in the satellite television market of the UK. Following the case discussion, we developed a simple game of war of attrition in the 'War of Attrition' section in Chapter 4. Refer to Figure 4.8 for the payoff matrix of the one-period game. The game gets repeated if the outcome is (stay, stay). It was discussed in the section 'War of Attrition' why the firms must either continue to stay till the others quit, or they must quit in the very first period. But neither actually happens in reality. Often, the firms continue to fight for some time and then either one of the firms quits or they stop fighting and share the market. Sometimes one firm takes over the other or the firms merge. This reality of war of attrition can be explained in terms of mixed strategies.

Suppose Firm B chooses to quit in the first quarter with probability p. That means Firm B chooses to stay till Firm A quits with probability $(1 - p)$. Now, if Firm A decides to quit in the first period, Firm A's expected payoff is $[5x \times p + 0 \times (1 - p)]$ $= 5xp$. On the other hand, if Firm A decides to stay till Firm B quits, then with probability p Firm A gets $9x$, but with probability $(1 - p)$, gets $-x$ in the first quarter and the game continues to the second quarter. Again, in the second quarter, Firm B chooses to quit with probability p. So, Firm A's expected payoff from quitting in the second quarter is again $5xp$, and if Firm A decides to stay till Firm B quits, then with probability p Firm A gets $9x$ in the second quarter but with probability $(1 - p)$, gets $-x$ in the second quarter and the game continues to the third quarter. That way the game can continue. So, if Firm A decides to stay till Firm B quits, then their expected payoff at the beginning of the first quarter is:

$$Z = 9x*p + (1-p)*[-x + \delta\{9xp + (1-p)(-x + \delta(9xp + ...)\}]$$
$$= 9xp + (1-p)[-x + \delta\{Z\}]$$
Or, $Z = [9xp - x(1-p)] / [1-\delta(1-p)]$

Since we are evaluating future payoffs at the beginning of the first quarter, the future payoffs are discounted. δ is the discounting factor.

Firm A will be indifferent to quitting in quarter 1 or deciding to stay till Firm B quits if and only if $5xp = [9xp - x(1 - p)] / [1 - \delta(1 - p)]$. For any given δ, this is a quadratic equation in p. Solving the quadratic equation, we will get the value of p for which Firm A becomes indifferent to quitting in quarter 1 or deciding to stay till Firm B quits. Suppose $\delta = 1$, that is, Firm A does not discount future payoffs. The equation becomes $5p^2 - 10p + 1 = 0$. Solving that equation, we get $p = 1.894$ or $p = 0.106$. p is a probability and must be a fraction. Therefore, $p = 0.106$. This means Firm A is indifferent to quitting in quarter 1 or deciding to stay till Firm B quits if and only if Firm B quits with probability 0.106. Since the game is symmetric between the two players, we may conclude that in a mixed strategy Nash equilibrium the firms will choose to quit in the first quarter, and in any quarter thereafter, with probability 0.106. Therefore, in the mixed strategy Nash equilibrium, each firm decides to stay till the other firm quits with probability 0.894. The game continues to the next quarter if both firms decide to stay, which means the game continues with probability $(0.894)^2 = 0.799$. In other words, there is close to 80 per cent chance that the war of attrition continues. If one firm has a reputation of not quitting, it induces the other firm to think that the rival firm will stay with a probability higher than that warranted by the mixed strategy Nash equilibrium. So, by developing such a reputation a firm can induce the rival firm to quit.

Games of Incomplete Information

Consider a game of poker in its simplest form. There are only three cards in the deck—the Ace, the King and the Queen of Hearts. Ace is the biggest card, followed by the King and the Queen, respectively. Two cards were dealt between Player 1 and Player 2 and the third card was kept aside. After the cards were dealt, the players could see their own card but not the card of the rival or the card that was kept aside. The deck was well shuffled and it is safe to assume that the rival player has got one of the other two cards with equal probabilities. Player 1 must first decide whether to bid or to fold. If Player 1 folds, she/he loses the game and pays ₹10 to Player 2. If Player 1 bids, Player 2 must decide whether to fold or call. If Player 2 folds, then he loses and in that case Player 1 gets ₹10 from Player 2. If Player 2 calls, that is, challenges Player 1, then the players must show their cards. Whoever had the bigger card wins the game. In this case, the winner gets ₹20 from the loser.

Suppose you are Player 1 and your card is the King of Hearts. What should you do? Bid or fold? Note that you have to make a decision under incomplete information. Player 2 might have the Ace, which is a bigger card than your card, or he might have the Queen. Both events are equally likely. How do you make a decision in a game like this?

Note that there are two possibilities—either Player 2 has the Ace, or he has the Queen. If he has the Ace, he will surely call as he will know that he has the biggest card. On the other hand, if he has the Queen, he will surely fold as he knows that he has the smallest card. Since Player 2 has the Ace with probability 0.5 and the Queen with probability 0.5, you may presume that you will lose ₹20 with probability 0.5 and get ₹10 with probability 0.5. So, your expected payoff from bidding is $[(-20) \times 0.5 + 10 \times 0.5]$ = −5. If you fold, then you lose ₹10, that is, your payoff is −10. Since your expected payoff is more from bidding, you should bid if you are risk-neutral.

This game is a game of incomplete information because Player 1 doesn't know what card Player 2 has. The way Player 2 will play if he has the Queen is very different from the way he will play if he has the Ace. Basically, there are two possible games depending on the card Player 2 has got, but Player 1 does not know which game he is playing. Hence, this game is a game of incomplete information. If Player 2 were the first mover, she could have bluffed out Player 1. Suppose Player 1 got the King and Player 2 got the Queen. Since the Queen is the smallest card, under complete information Player 2 would never bid. But since Player 1 doesn't know whether Player 2 has the Queen or the Ace, if Player 2 bids, Player 1 may think that she has the Ace and hence Player 1 might fold. That is bluffing. Riding on the incompleteness of information, it is possible to bluff. Indeed there is a risk involved. Player 2 also does not know whether Player 1 has the King or the Ace. If Player 1 had the Ace, he would have known that his card is the biggest one and in that case he would have challenged Player 2's bid. Even with the King, Player 1 can challenge Player 2's bid, but in that case he takes a risk.

Bluffing is an integral part of poker. It is possible to bluff in a game of poker because the other players don't know what cards you have. In other words, poker players don't know enough about their rivals. In the rest of this chapter, we will discuss games of incomplete information where the players might not know their rivals well, that is, they don't know the 'types' of the other players in the game. In the absence of knowledge about 'types' of other players, we make decisions on basis of our 'beliefs' about the 'types' of the other players.

Player Types and Belief

Consider a situation wherein you are coming out of a bank with a large amount of cash when you felt a pistol on your ribs. It's broad daylight on a crowded street in downtown. The man behind you is asking for the cash and is threatening that he will shoot you otherwise. Suppose you are very cool and is capable of thinking rationally even in a situation like this. So, you are able to apply your understanding of game theory and possibly outsmart the extortionist. As it is customary in game theory, assume that the extortionist is also rational. Now, if the extortionist thinks he won't be able to escape with the money after shooting you, then his threat is a non-credible one. You can look forward and reason backward. Suppose we have complete and perfect information, and by virtue of that complete information about the man you know that he won't be able to escape after shooting you, and the man also knows that. So, looking forward you figure out that if he shoots he will be caught and will get a life sentence. Assuming that he is rational, reasoning backward you can figure out that he won't shoot even if you don't part with the money. Hence, in a world of complete and perfect information, being rational you won't part with the money. Unfortunately, we don't have complete information. If the man thinks that after he shoots, everyone around will get scared and run after their lives, and before the police comes he will be able to escape with the money, he might shoot. Note that we are still assuming that everyone involved here is rational. But you don't know whether the extortionist thinks he will be able to flee with the cash or he doesn't. That means you don't know the 'type' of the man behind you. If he is very optimistic about being able to escape, he might shoot, and in that case your best response is to part with the money since you value your life much more than the money in question. On the other hand, if he is more realistic and thinks that the police van stationed within hundred metres will be able to catch him, he won't shoot, and in that case your best response is to refuse to part with the money. So, your knowledge of game theory would have been helpful if you knew the type of the extortionist. But you don't know his type, that is, you have to make decision with incomplete information about the type of the other player. The situation is like a poker game.

Is game theory useless in the world of incomplete information? No. It can still help you make a decision. But your decision will depend on your 'belief', and also on your risk appetite. Mathematically speaking, 'belief' is a probability distribution over the set of 'types' of your rival player. Here, in the mentioned example, suppose there are two possible types of the extortionist

Figure 7.3

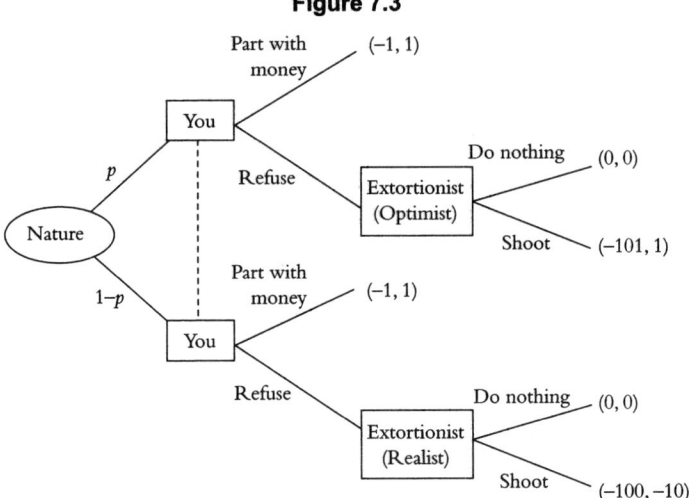

as stated—the 'optimist type' who thinks he will be able to escape with the money after shooting, and the 'realist' type who thinks that he will be caught by the police if he shoots. Your belief is what you think is the chance that the extortionist is the 'optimist type'. We incorporate belief in a game tree by introducing another player called 'nature'. 'Nature' does not have any stake in the game, but only makes a random draw of the player's type from the set of types. The game tree of the extortion game is given in Figure 7.3.

Suppose the amount of money is ₹1 million and value of your life to you is ₹100 million. In case the extortionist shoots and gets caught, he should get a life sentence. Let the value of his life to him is ₹10 million.

First nature draws the type of the extortionist and hence the game. Nature draws the 'optimist type' with probability p and hence the 'realist type' with probability $(1 - p)$. Nature is basically reflecting your own belief. Post nature's move, the two branches represent two different games. The upper branch represents the game with the extortionist if he is an 'optimist', and the lower branch represents the game with the extortionist if he is a 'realist'.

Since you don't know the extortionist's type, you don't know which game came up for you in the random draw by nature. But you believe that you are put up against an 'optimist' extortionist with probability p and against a realist extortionist with probability $(1 - p)$. The fact that you don't know which game you are in is represented by the broken line joining the nodes you move in the two different branches. Technically, it is called an information set, and the information set represents your decision node. How do you decide in such a situation?

In order to see how you will make a decision, we need to know the exact probability distribution that represents your belief. Suppose you think that there is a 10 per cent chance that the extortionist is an 'optimist'. That means, you believe that he is an 'optimist' with probability 0.1 and a 'realist' with probability 0.9, that is, nature has chosen $p = 0.1$. Let us also assume that you are risk-neutral, that is, you maximize your expected payoff. If you part with the money, you are losing ₹1 million. But if you refuse to give in, then you are taking a chance. If the extortionist is actually an 'optimist', he will shoot and you will lose your life as well as the money. But if the extortionist is a 'realist', he won't shoot and you will lose nothing. So, if you refuse to give in, with probability 0.1 you will lose ₹101 million (value of life + money) and with probability 0.9 you lose nothing. The expected payoff from refusing is $[0.1 \times (-101) + 0.9 \times 0] = -10.1$. Since $-10.1 < -1$, you will better part with the money. We can actually find the critical minimum value of p for which you will part with the money. When you believe that the extortionist is an 'optimist' with probability p, you will part with the money if $-1 \geq -101p$, that is, if $p \leq 1/101$. This means you will part with the money if you believe that the extortionist is an 'optimist' with a chance larger than or equal to 1 in 101. If you are risk-averse you will part with the money if the chance of the extortionist being an 'optimist' is even less than 1 in 101. Constructing the situation as a game of incomplete information helped us to understand why even rational people will better part with the money even knowing that there is a very low chance that the extortionist will be able to flee with the money after shooting.

Now let us explore a few business scenarios that could be analysed as games of incomplete information. We will explore the strategic issues of reputation building and signalling.

Reputation

Reputation is unimportant in one-shot games. But in games that are played repeatedly, either against the same player or against different players which is observed by all, reputation acts as a mechanism to influence beliefs of rivals. In a way, reputation acts as a commitment. Suppose a terrorist outfit abducts a random person and using that person as a hostage tries to negotiate for release of some of their leaders. The outfit knew that the government may or may not give in to their demand. They might believe that the two possibilities are equally likely. If the government does not give in and refuses to negotiate, it develops a reputation of not negotiating with terrorists. The cost of this reputation building is loss of some precious lives. But next time when that very outfit or another similar terrorist outfit attempts to negotiate with the government holding someone as hostage, they would update their belief based on the past experience. They would think that the probability that the government would agree to negotiate is rather low. The recurring refusal by the government would affirm the government's reputation and in future, terrorist outfits will know that their attempts to negotiate using civilians as hostages are futile. In order to abduct a civilian of some importance and keep him or her as a hostage, the terrorist outfits incur costs in terms of effort, resources and risk. They won't want to bear that cost if they know that the probability of their success is very low. Of course, the government and the civilians pay a price for this in terms of lost lives. The strategy of diffusing terrorist attempts by refusing to negotiate works if it is backed up by a strong state machinery. If abducting someone and holding them as hostages are easy for the terrorists, they will not mind bearing that cost even if they know that the probability of their success in negotiating with the government is very low.

The reason why some retail outlets maintain a 'fixed price' policy could be explained in similar terms. The refusal to bargain and holding on to that commitment builds their reputation of not offering any customer-specific discount. In a bargain shop, the customer never knows whether she/he has got the best bargain. The customers who think they cannot bargain well prefer a 'fixed price' shop. So, having the reputation of refusing to bargain self-selects the customers who think they cannot bargain well and saves efforts of the sales personnel. There might be some loss of revenue for the 'fixed price' outlets. Reputation generally comes at a cost.

Sometimes acting irrational gives you a reputation of being somewhat crazy, and that reputation helps!

Predatory Pricing

Predatory pricing refers to selling below cost with the intent to kill competition. It can be done either to force the competitor to exit the market, or to deter entry of potential entrants. In most countries antitrust laws classify predatory pricing as anti-competitive and, hence, illegal. But it is difficult to prove the intent of predation. Typically, the state or a competitor files an antitrust case alleging predatory pricing. But the plaintiff firm can defend themselves because it is possible for firms to fudge their cost data. The plaintiff argues that they are just being competitive and the low price is beneficial for the customers. The argument against predatory pricing is that it weeds out competition in the long run and the firm gains monopoly power, which is not good news for the consumers. Doing so, the predator loses money. But losing money in the short run is alright if the loss can be more than compensated in the longer run. Case Study 7.1 shows a few instances where Walmart had been accused of precisely the same kind of predatory pricing.

There had been numerous instances, like the ones mentioned in Case Study 7.1, when Walmart had been accused of predatory pricing. It was always argued that in order to recover the loss, the retail giant intended to raise price in future. Indeed, the predator

Case Study 7.1: Predatory Pricing and Walmart

In 1997, Walmart entered Germany by acquiring two unprofitable retail chains—Wertkauf and Interspar. These two chains had only 3 per cent share in the German retail market. In order to compete against Metro AG and Rewe Group, the two largest retail chains in Germany, Walmart emulated its US market strategy known as "Everyday low prices". Metro AG and Rewe Group responded by lowering prices. In September 2000, the Federal Cartel Office accused Walmart and its two biggest German rivals of selling about a dozen staple products like milk, butter, vegetable oil, etc., at prices below respective average costs. A fine of one million deutsche mark was slapped on Walmart. The products in question are what Walmart classified as "corner products". As a strategy Walmart used to price these corner products very low everywhere. As per Walmart's understanding, the consumers know the prices of these items with other retailers, and if these items are priced low it creates a general impression of low price with the consumer. Indeed the slogan "Save Money Live Better" made Walmart the largest retailer in the USA. Walmart appealed against the accusation of predatory pricing. But the Supreme Court of Germany ruled in January 2003 that Walmart violated the country's anti-trust laws with their below-cost pricing strategy. The ruling stated that the authorities feared that a price war among the big three retailers will decimate the mom-n-pop stores leaving the consumers with fewer choices.

Around the same time Crest Foods, a three-store chain of food items retailer in Oklahoma City in the USA filed a predatory pricing suit against Walmart. Edmond Walmart Supercentre opened in Oklahoma City on 18th May 2000. Allegedly, on 23rd May 2000 a team of Walmart personnel including former president Mr David Glass visited Crest Food and scanned their prices using a hand-held device. Later, during their investigation, law firm Crowe and Dunlevy found that Edmond Walmart Supercentre was beating Crest in prices of 25 items, while Crest was beating other Walmart

(Case Study contd.)

(Case Study contd.)

supercentres in 22 items. Edmond Walmart Supercentre was beating other Walmart supercentres in prices of 28 items. It was cited as an example that on 9th June 2000, French Mustard sold for 88 cents in Crest, while in Edmond Walmart Supercentre the same item sold for 50 cents. However, on that very day the same item was priced in the range of 88 cents to 97 cents in other Walmart supercentres in Oklahoma City and nearby areas. Crowe and Dunlevy concluded that Walmart's pricing resulted in 20 per cent sales drop for Crest Foods, which amounted to a damage of 3 million dollars in a span of three months.

While the trouble was brewing for Walmart in Oklahoma, Wisconsin Department of Agriculture, Trade and Consumer Protection charged Walmart with selling staple goods like milk, butter, detergent, etc., at prices below cost in several stores within the state of Wisconsin, which violated the state's antitrust laws. The complaint stated that Walmart intended to force other stores out of business, gain monopoly in local markets and ultimately recoup its losses through higher prices in future.

Sources: Stacy Mitchell, "German High Court Convicts Wal-Mart of Predatory Pricing", 2003, available at http://www.ilsr.org/german-high-court-convicts-walmart-predatory-pricing/ (last accessed 28 July 2015).

Gregory Potts, "Crest Sues Wal-Mart Over Edmond Store Pricing", 2000, available at http://newsok.com/crest-sues-wal-mart-over-edmond-stores-pricing/article/2713349 (last accessed 28 July 2015).

must raise price after competition is eliminated. But there is a flaw in this argument. Suppose a large firm successfully eliminates competition by pricing products below cost. When the price is increased after competition is eliminated, new competitors will enter the market. So, the below average cost pricing must be continued in order to retain monopoly position. If a firm continues with below average cost pricing for good, it continues to make losses despite being a monopoly. That is absurd. But predatory pricing, defined as below average cost pricing with the intent of elimination of competition, has been proved in quite a few instances. What is it that motivated predatory pricing then? Pricing below average cost to impose exit on competitors makes sense only if the predatory pricing deters entry of new competitors as well. It is like killing two birds with one stone. Predatory pricing might actually deter entry of new competitors in future due to the predator firm's reputation that they will price below the average cost whenever they face competition. The predator may not gain that reputation just because they eliminated competition by predatory pricing once. Nevertheless, it influences the belief of potential entrants. A consistent history of predatory pricing by a firm establishes its reputation. Let us see how it happens with the help of a simple example.

Consider a two-period game between the predator firm (Firm P) and two of its rivals, Firm R1 and Firm R2. Actually R1 and R2 might be the same firm, or they might be different organizations. For clarity of our understanding, we will consider them as two different players at two different time periods. In period-1 of the game, Firm P decides whether to prey on Firm R1 or not. If it decides to prey, it prices below its average cost in period-1 and there is a price war. Both firms lose money in period-1, and Firm R1 exits at the end of period-1. Of course, R1 might not exit at the end of period-1. In that case, P fails to impose exit on R1, that is, the fundamental purpose of predatory pricing fails. If Firm P doesn't prey by pricing below cost in period-1, then both firms charge competitive prices and make profits in that period. Suppose each firm earns 100 under competitive scenario. But if there is a price war, each firm loses 50.

Irrespective of whether Firm P preyed on Firm R1 in period-1, Firm R2 decides whether to enter or not at the beginning of period-2. If Firm R2 enters, Firm P may again fight by pricing below average cost and incurring losses. It may also accommodate the entry of Firm R2 by pricing its product competitively without any intent of predation. If there is entry and the entry is accommodated, each firm earns 100. If entry of Firm R2 is fought by Firm P, there is a price war. Price war results in a loss of 50 for each firm in period-2 as well. If there is no entry in period-2, Firm P earns the monopoly profit in that period. Let the monopoly profit be 500. In the context of our game here, R1 and R2 are the same organizations if R1 does not exit at the end of period-1. So we will conclude that predatory pricing works if and only if we find that R2 does not enter in period-2.

In order to keep the game simple and tractable, we restrict the game to two periods and assume that the game ends at the end of period-2. In reality it may go on. There may be many periods following, and in each period a new entrant decides to enter or not, and Firm P must decide whether to accommodate entry or not. So long as the number of periods is finite, we can solve the multistage game using backward induction. In principle, there should not be any fundamental difference in the result depending on the number of periods in the game. So, the two-period model should work fine. The game tree of the two-period game under complete information is given in Figure 7.4.

R1 does not have any active decision node in Figure 7.4. If P preys, R1 incurs loss in period-1 and exits. If P does not prey, R1 gets 100 in period-1. R1 is not there in period-2. As mentioned earlier, R1 and R2 may be the same firm though. The payoffs are

Figure 7.4

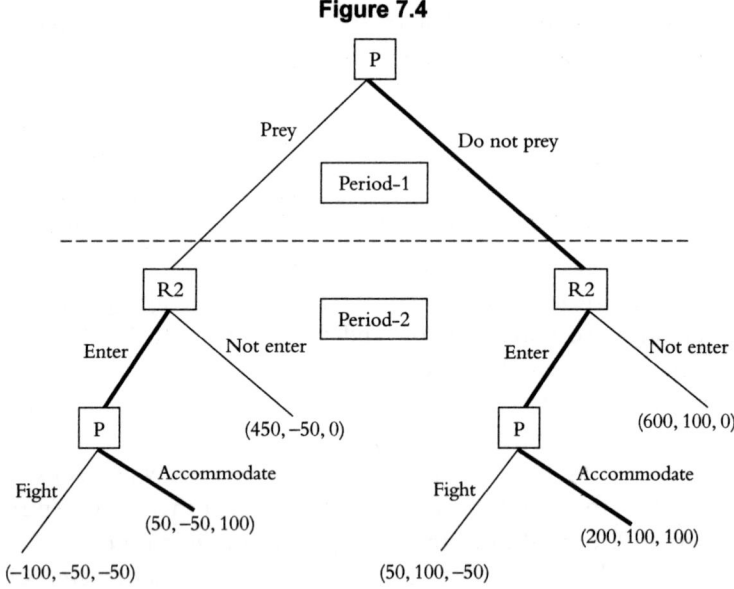

written as (P's payoff, R1's payoff, R2's payoff). For P, the payoff is the sum of payoffs from two periods. If this game is played under complete information, we should be able to solve it by method of backward induction. In Figure 7.4, the bolded branches indicate optimal decision at the respective nodes. In period-2, P does not have any reason to fight. R2 will be able to foresee that its entry will be accommodated. So in period-2, R2 will surely enter, irrespective of whether P preys on R1 in period-1. P should be able to foresee that predation in period-1 won't pay. Since R2 will enter in period-2 in any case, P is better off not preying on R1 in period-1.

The above example shows the flaw of predatory pricing. Indeed, predatory pricing is mindless if there is complete information. The problem was first identified by Nobel laureate German economist Reinhard Selten. In his seminal article titled "The chain store paradox," Selten recognized the futility of predatory pricing. He called it a paradox because in reality chain stores like Walmart often indulge in predatory pricing. In fact, predatory pricing might be a smart strategic move in a world of incomplete information. After all, Firm P might not be making rational decisions, and it might be predisposed to fight any competitor. In other words, R2 does not know the type of P and makes decision like a poker player. Let's rework the game under incomplete information.

Suppose R2 believes that P might be either type-1 or type-2, and that P is type-1 with a probability ρ. Type-1 is rational and makes decisions on basis of payoff calculations. But type-2 is predisposed to fighting. There might be some sort of rationality behind such predisposition too. If P is type-2, they might think that continuous preying of competitors will make them a global monopoly in the long run. It requires deep pockets to execute such a strategy. At least in the immediate future it looks like a crazy option. With this modification, the period-2 game looks like the one given in Figure 7.5.

Here, we have taken only the period-2 payoffs for P and R2. Since R2 moves first, we have written the payoffs as (payoff of R2, payoff of P).

Figure 7.5

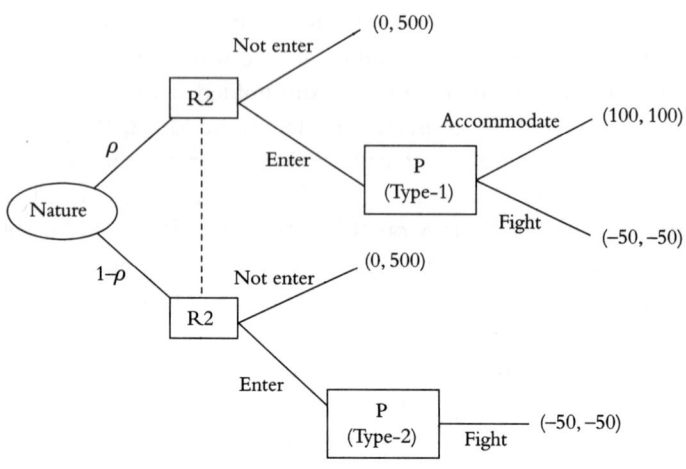

R2 believes that P is type-1 with probability ρ and type-2 with probability $(1 - \rho)$. In the game tree, we represent R2's belief as nature's draw. Without knowing whether P is type-1 or type-2, R2 must decide whether to enter or not. If P is type-1, it will decide to accommodate the entry, as P's payoff is more when the entry is accommodated. But if P is type-2, it will fight. For type-2, accommodation is not an option. So, the expected payoff of R2 from entering is $[\rho \times 100 + (1 - \rho) \times (-50)] = [150\rho - 50]$. Therefore, R2 will enter if $[150\rho - 50] > 0$, that is, $\rho > 1/3$. In other words, R2 will enter if it believes that P is type-2 with probability less than or equal to 2/3. But if R2 believes that P is type-2 with a probability more than 2/3, then it will not enter, and in that case P becomes a monopoly in period-2.

Predatory pricing to impose exit on R1 in period-1 influences the belief of R2. Without any history of P preying on its rivals, R2 might believe that the probability of P being type-2 is low. But if R2 knows that P preyed on R1 in period-1, then R2 will update its belief. One-off instance might not influence the belief significantly. But if there is a consistent record of P preying on its rivals, then R2 might believe that P is type-2 with a sufficiently high probability, and hence will not enter. It is like a game of poker in which you consistently bid or raise with low cards to make your rivals believe that you are predisposed to raising. You will lose for a few rounds, but after gaining the reputation, you may raise on a large pot and get a call from your rival which might make your fortune.

In a drastic attempt to wipe out all competitors, an entrant with a deep pocket may use predatory pricing in all markets that it operates in. The Ola taxi hailing service in India is a very contemporary example closer home.

Ola operates like a chain-store in the sense that they have presence not only in all the Indian metros, but also in smaller towns. Particularly in the smaller towns there is no app-based rival of Ola. In these towns, Ola's only rivals are local cabs and auto-rickshaws. If these small entrepreneurs go out of business, Ola will have a monopoly position in these towns. As the Ola

---

Case Study 7.2: Ola's Alleged Predatory Pricing

Smartphone app-based taxi hailing service is still at a nascent state in India. The market was created by USA-based Uber Technologies. ANI Technologies, which runs Ola, entered the market later but took it over by their low fares for passengers and high incentives for drivers. As this book goes to press, Ola, India's largest taxi hailing service is being probed for predatory pricing by Competition Commission of India (CCI). Fast Track Call Cab, a rival of ANI Technologies, filed a complaint with CCI that Ola is using predatory pricing in Bangalore. Primary investigation of CCI shows that Ola is spending ₹574 per trip, on an average, in Bangalore, while their revenue per trip is ₹344. That means, Ola is spending ₹230 per trip to kill competition. A large number of well-funded companies, including Japan's SoftBank, invested in Ola. Allegation is that Ola is using the funds to drive out equally efficient competitors who cannot match such pricing due to lack of resources. The complainant also charged Ola with creating barriers for new entrants.

Source: Ribeiro, 2015.

fleet keeps increasing, the drivers might find work through Ola, which also will help the cab hailing service to keep yellow cabs and auto-rickshaws out of market. So, Ola's objective may not be just elimination of rivals in cities like Bangalore, but also to create a reputation of being hyper competitive, which in turn helps them in deterring the entry of new entrants in future.

Signalling

Signalling refers to the mechanism through which a player, whose type is not known to other players, sends a message to them either to reveal his or her type or to mislead them about his or her type. The player who sends the message is called the 'sender' and the ones who receive the message are called 'receivers'. The message sent by the sender does not necessarily reveal the sender's type. Observing the message, but without knowing the type of the sender, the receiver must make a decision.

Bridge, which is a game of cards, is a signalling game. In a game of bridge, four players make two teams of two each. The four players sit on four sides of a table with the members of the same team sitting facing each other. The players are denoted as North, East, South and West. While North and South make a team, East and West make the rival team. In competitive bridge games, a screen is kept between the partners to prevent them from making eye contact. This is done to prevent the players from making gestures to signal. Bridge is a signalling game, but signals must be made with calls. The entire deck of 52 cards is shuffled and dealt equally to four players. Each player can see their own cards but not of the others. The game is divided into two phases—the auction phase and the playing phase. During the auction phase, the players make calls to make an agreement about who takes how many tricks and for establishing which suit, if any, will be trumps for an undertaking to win the specified number of tricks. During this phase the players try to reveal to their respective partners what kind of cards they have and they also try to mislead the rivals. Knowing your partner's call and the rivals calls, but not knowing exactly what cards they have, you must make calls. Some initial calls are made to send across messages to the partner. The side winning the auction is called the declaring side and the other side is known as the defending side. During the playing phase, the dummy hand of the declaring side is laid down. So, during this phase each player can see 26 cards. The memory of the calls holds importance in this phase too, and the defending side can still make signals to each other through their game play.

There are a lot of business scenarios that are similar to the game of bridge in the usage of signalling mechanisms. In the section 'Reputation', we discussed predatory pricing and explained it using reputation. Another explanation can be provided using signalling. Increasing the debt-to-equity ratio may act as a signal of insider confidence to investors. Quality or process certification acts as a signal of service quality for firms in service sector. Higher education does not necessarily impart students with skills required for the job market. Still people acquire higher education to signal their ability in the job market. In this section, we will discuss some of these business applications of signalling.

Predatory Pricing as a Signal of Cost

Signalling provides an alternative explanation of predatory pricing. This signalling-based argument that justifies predatory pricing also explains the chain-store paradox. A firm that operates in multiple geographic locations, faced with entry of a rival in one market and threatened by the potential entry of rival firms in other markets that it operates in, tries to deter potential entrants by the use of predatory pricing in the market where entry occurred. Folgers coffee, a brand owned by Proctor & Gamble, and Maxwell House Coffee, a brand owned by Kraft foods used to be the two leading brands of coffee in the USA during the 1990s. Before a coffee war broke out between these two giants, they avoided a war of attrition for many years by following a simple rule of conflict resolution—geographic dominance. Folgers used to operate in the region spanning from the west bank of the river Mississippi to the Pacific coast. Maxwell

House used to dominate the region between the east coast to the east bank of Mississippi. The coffee war broke out when Folgers entered Cleveland, which used to be within the area of Maxwell House Coffee. In retaliation, Maxwell House Coffee started selling at a very low price that was alleged to be below their average cost. Maxwell House Coffee didn't sell at that low price only in Cleveland, but also in Pittsburgh and Syracuse. This behaviour can be explained using the limit pricing argument.

Limit price is the price at which the entrant makes losses but the incumbent does not necessarily make losses. If the incumbent firm is a low-cost firm, it does not incur losses when they set price low enough to ensure that the potential entrant makes losses if they enter. However, if the incumbent is a high-cost firm, then it makes losses at that price. The entrant doesn't know if the incumbent firm is a high-cost firm or a low-cost firm, that is, they don't know the type of the incumbent. The incumbent sends a message about its cost through its price. The potential entrant only observes the price and decides whether to enter or not. Consistent predatory pricing makes the potential entrants believe that the predator is a low-cost firm. At least, they revise their beliefs and presume that the probability of the predator being a low-cost firm is very high. In this case, if the predator is not a low-cost firm, it successfully bluffs out the potential entrants through false signalling.

Let us try to understand the nuances of the signalling game with the help of an example. Suppose Firm A operates in two markets—Market-1 and Market-2. Firm B entered Market-1 and is contemplating future entry in Market-2. After Firm B entered Market-1, Firm A can either charge an entry-accommodating competitive price, or it might charge a limit price that is lower than the competitive price. If Firm A charges the entry-accommodating price, post entry Firm B earns a profit of 10 million dollars from Market-1. However, if Firm A charges the limit price, Firm B makes losses of 40 million dollars. These profits (or losses, as the case may be) could be thought of as the PDV of future profits (or losses) over a finite time horizon. The catch here is that the incumbent, that is, Firm A, can be either a low-cost firm or a high-cost firm. If Firm A is a high-cost firm, then at the limit price they make losses. This means their average cost is higher than the limit price. Firm A, if it is a high-cost firm, also loses 40 million if they choose the limit price in Market-1. Instead, if they charge the entry-accommodating price, then they make a profit of 50 million. But if Firm A is a low-cost firm, that is, their average cost is less than the limit price, then they make a profit of 80 million even when they choose the limit price. Rather, if they charge the entry-accommodating price, their profit reduces to 60 million. Note that even though the entry-accommodating price is higher, the market gets shared at that price and in turn results in reduced profit. The monopoly profit of Firm A in the Market-1 is 500 million if it is a high-cost firm, and is 510 million if it is a low-cost firm. Let us assume that the profit (or loss) figures in Market-2 are identical to those in Market-1 under same conditions.

If Firm B knew that Firm A is a high-cost firm, it would have been certain that the entry would be accommodated and it would earn 10 million from each market that it enters. Since Firm A, if it is a high-cost firm, loses 40 million in each market from limit pricing, it does not make economic sense for it to choose limit price. On the other hand, if Firm B knew that Firm A is a low-cost firm, it would have known that its entry would be fought by limit pricing since a low-cost incumbent earns more at the limit price than at the entry-accommodating price. So, if Firm B knew that Firm A is a low-cost firm, it would have stayed out. Here, in this example, Firm B does not know whether Firm A's average cost is more than the limit price or less, that is, Firm B does not know the type of Firm A. Firm B makes entry decision under incomplete information on the basis of their prior belief.

After Firm B entered Market-1, Firm A will surely charge the limit price if it is a low-cost firm. But if Firm A is a high-cost firm, apparent wisdom is in accommodating entry. If it charges the limit price, they incur a loss of 40 million in Market-1. Instead, if it accommodates entry, it makes a profit of 50 million. On the flip side, if Firm A accommodates entry of Firm B in Market-1, Firm B will be sure that Firm A is a high-cost firm and would enter Market-2 as well. By accommodating entry in both markets, Firm A gets an aggregate payoff of 100 million if it is a high-cost firm.

But it might be possible for Firm A to outsmart Firm B by bluffing. Even if Firm A is a high-cost firm, using its post-entry pricing decision in Market-1 it might be able to convince Firm B that it is a low-cost firm. It loses 40 million in Market-1 due to limit pricing, but by doing so if it can keep Firm B out of Market-2, then its aggregate payoff from the two markets becomes 460 million dollars, which is more than its aggregate profit from accommodating entry in both markets. Now, using a structured signalling game, let us figure out the conditions under which it is really possible for Firm A to outsmart Firm B in that manner.

As is done in games of incomplete information, we represent Firm B's belief as a random draw by 'nature'. Firm A is the 'sender' and Firm B is the 'receiver'. Firm A uses its pricing decision in Market-1 as a message to signal its type to Firm B. Firm B observes this message, but not the type, and decides whether to enter Market-2 or not. The game tree of the signalling game is given in Figure 7.6. The decision of Firm B to enter Market-1 is not shown in the game tree. That decision prompted this signalling game. The payoffs are written as (payoff of Firm A, payoff of

Figure 7.6

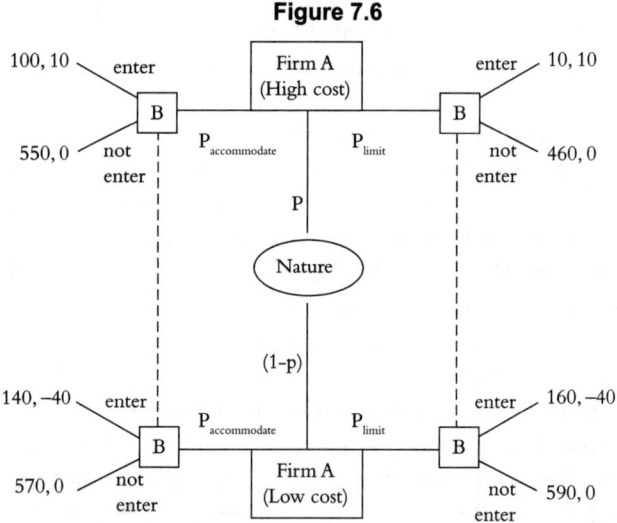

Firm B). In any signalling game, the sender is the first mover and hence the sender's payoff is written first. Note that the payoffs of Firm A are aggregate payoffs from two markets. But for Firm B, we have taken only the payoff from Market-2. In this game, Firm A's decision in Market-1 affects its aggregate payoff. But Firm B took the decision to enter Market-1 before the signalling game began. Firm B's payoff from Market-1 is dependent on the action of Firm A, and Firm B's experience in Market-1 might affect its decision to enter Market-2. But Firm B does not optimize on its aggregate payoff while deciding to enter Market-2. If Firm B enters Market-2, Firm A will accommodate the entry if Firm A is a high-cost type, and will choose limit price if it is a low-cost type. Since there are only two markets, there is no reason for the high-cost type Firm A to limit price in Market-2.

A priori Firm B believes that Firm A is a high-cost firm with probability p. This belief is formed on the basis of historical data on the proportion of high-cost firms in comparable industries and is same as the belief under which Firm B entered Market-1. Since the prior belief is formed on the basis of historical data, both Firm B and Firm A know the probability distribution.

After Firm B enters Market-1, Firm A can either charge $P_{\text{accommodate}}$ or P_{limit}. $P_{\text{accommodate}}$ is the entry accommodating price and P_{limit} is the limit price. Note that Firm A will not charge $P_{\text{accommodate}}$ if it is a low-cost firm. So there are two possibilities—Firm A charges P_{limit} irrespective of whether it is a low-cost type or high-cost type, and Firm A charges $P_{\text{accommodate}}$ if it is high-cost type and charges P_{limit} if it is low-cost type. In the former case, Firm A's message does not reveal its type, and in the latter case, the message reveals its type. When the sender chooses the same message irrespective of its type, we say that the sender chose a 'pooling strategy'. When the sender's message gives away its type, we say that the sender chose a 'separating strategy'. The receiver observes only the message sent by the sender, but not the sender's type. Here, in our example, Firm B is the receiver and they only observe whether Firm A chose $P_{\text{accommodate}}$ or P_{limit} in Market-1. When Firm B observes $P_{\text{accommodate}}$, they conclude with certainty that Firm A is a high-cost firm and decides to enter Market-2. In that case, Firm A gets 100 in aggregate and Firm B gets 10. Question is whether it is viable for Firm A to choose P_{limit} even if it is a high-cost firm. When Firm B observes P_{limit}, they cannot infer the type of Firm A. If Firm B conjectures that Firm A chooses $P_{accommodate}$ if Firm A is a high-cost firm and chooses P_{limit} if they are a low-cost firm, then Firm B enters Market-2 when they observe $P_{accommodate}$ in Market-1 and stays out when they observe P_{limit}. Given their conjecture, Firm B's decision is the best response, but the conjecture is not rational. If Firm B stays out whenever they observe P_{limit}, then Firm A should choose P_{limit} irrespective of their type. The alternative conjecture of Firm B is that Firm A used a pooling strategy. Given this conjecture, Firm B must decide on entry in Market-2 on the basis of their prior belief. Since Firm B believes that Firm A is a high-cost type with probability p, they will enter Market-2 if and only if [p × 10 + (1 − p) × (−40)] > 0, that is, if and only if $p > 4/5$ and stay out if $p \leq 4/5$. In other words, upon observing P_{limit} in Market-1 if Firm B conjectures that Firm A used a pooling strategy, Firm B's best response is to enter if they believe that there is less than one in five chance of Firm A being the low-cost type and their best response is to stay out if they believe that the chance of Firm A being the low-cost type is more than one in five. The conjecture that Firm A used a pooling strategy is not rational under the belief that $p > 4/5$. Since Firm A is aware of Firm B's prior belief, when $p > 4/5$ Firm A will not choose P_{limit} in Market-1 if it is a high-cost firm. However, the conjecture that Firm A used a pooling strategy is rational under the belief that $p \leq 4/5$. When $p \leq 4/5$, Firm A will choose P_{limit} in Market-1, irrespective of whether they are a high-cost firm or a low-cost firm. Riding on Firm B's belief that the incumbent firm is a high-cost firm with a probability less than 4/5, Firm A can bluff Firm B by choosing P_{limit} in Market-1 despite being a high-cost type. This way Firm A can outsmart Firm B and retain monopoly in Market-2. If we take the analysis one step backward, Firm B should be able to foresee that if $p \leq 4/5$, Firm A will use limit pricing irrespective of their type. Firm B should not enter even Market-1 if $p \leq 4/5$.

Higher Education as a Signal of Ability

Higher education is more about job market signalling than actual skill formation. Often the kind of skill higher education imparts to the students is not useful on the job. Then why do recruiters insist on certain qualifications? A recruiter may screen the candidates from the population and recruit suitable employees. For recruiting administrators, the Government of India conducts civil services examinations, which is a screening process. But the cost of such screening is huge and the process is unviable unless the recruiter wants to recruit in large numbers. Higher education screens out the candidates and reduces the size of the set to choose from. Let us take MBA education as an example. The MBA curriculum imparts the students with certain fundamentals, but management in practice is much more than what could be taught in the classroom. Indeed, all MBAs don't make successful managers. On the other hand, it is possible to become a successful manager without MBA. Nevertheless, most MBAs make reasonably efficient managers and it is generally a much longer process to reach the middle management level without an MBA. Why should that happen if MBA education does not have much to do with formation of practical management skills? It is because MBA acts as almost noiseless signal in the managerial job market. The process of MBA education does the trick. It is very competitive and extremely difficult to get admission in a top-level business school almost anywhere in the world. It is more competitive in India due to the existence of a large middle class and relatively smaller number of seats in the best schools of business in the country. The admission process of these schools ensures that they select the most efficient ones. Theoretically, it might be possible for any graduate to get admission in one of the best schools of business if they don't have any constraint of time and effort. Less efficient a student is, more difficult it is for him/her to get admission. Even if a student gets through by fluke, the rigour of the program ensures that such students get weeded out during the course of two years. In other words, even though in principle it might be possible for any graduate to become an MBA, the cost of effort required for the not so bright ones might be prohibitively high. Note that the ability to pay the fees should not be a deterrent in the presence of a perfectly functioning credit market. The students who get admission in the top schools get loans easily using the admission itself as collateral. Nobel laureate economist Michael Spence developed the job market signalling model. The crux of the model is that if the effort cost of obtaining higher education is sufficiently high for the less efficient individuals, and if the probability of a random individual being efficient is low, then it is not possible for the less efficient individuals to camouflage as efficient ones by obtaining higher qualification like them.

Consider a signalling game between a candidate and a recruiter. The candidate is the sender and the recruiter is the receiver. The candidate may be type-1, that is, efficient or type-2, that is, less efficient. To be precise, a type-1 candidate creates a higher value for the employer, if employed. The type of the candidate was drawn by nature. MBA is not essential for the job in question. The recruiter wants to select a type-1 candidate and reject a type-2. After graduating from college, the candidate decided to do an MBA to send a message about his type. The recruiter can observe the message in form of the degree, but does not know the candidate's type. The recruiter can select the candidate either in a managerial role with a higher pay, or in a non-managerial role with a lower pay. If the recruiter knew that the candidate is type-1, she/he would have recruited the candidate in a managerial role and otherwise in a non-managerial role. Nevertheless, MBAs are paid higher than people without MBA. For now, think of that as an MBA premium. After we solve the signalling game, we will understand why that premium is paid to MBAs.

Let us consider the following payoff landscape. The type-1 candidate, if employed, creates a value of 1000 for the employer, whereas type-2 creates 400. If the candidate is offered a managerial role, he will be paid 800, and if he is offered a non-managerial role he will be paid 300. These numbers are rupees in lakhs, which are the PDVs of streams of payoffs created and earned over a finite time horizon. During this period, the recruit goes through stints in various management functions, and at the end of the period is ready to join the middle management provided he gets the required performance appraisal. After that, it is a different game in the professional world. The purpose of the job market signalling is to get the first break. The candidate was aware of these payoffs when he decided to join the MBA programme. He was also aware that people without MBA get 600, if they are recruited in a managerial role and 200 in a non-managerial role.

The minimum offer acceptable to the candidate is his reservation payoff, which is dependent on the candidate's type. The reservation payoff of the candidate is higher if he is type-1. The candidate knows his type and hence knows his reservation payoff. The reservation payoff reflects the candidate's self-confidence. The recruiter doesn't know the type of the candidate and the candidate's reservation payoff is unknown to her. Suppose the reservation payoff of the type-1 candidate is 500 and that of type-2 is 200. This means if the candidate is type-1, he will accept only an offer of managerial role. But type-2 will accept any of the two roles offered by the recruiter. When the recruiter fails to recruit, her payoff is zero. Let x be the effort cost of obtaining an MBA for a type-1 candidate and $(x + \theta)$ be that for type-2. Since the cost of effort is more for type-2, θ is more than zero. The extensive form representation of the job market signalling game is given in Figure 7.7.

R is the recruiter. Nature's draw reflects the recruiter's belief. The recruiter believes that the candidate is type-1 with a probability 0.2 and type-2 with a probability 0.8. The basis of this belief is historical data on the proportion of type-1 candidates in the population of graduates. As an example we assumed that 20 per cent of the graduates are type-1. The belief of the recruiter is common knowledge, that is, the candidate is aware of the recruiter's belief.

Recruiter observed that the candidate is an MBA. But having MBA does not necessarily mean that the candidate is type-1. The recruiter does not know the type of the candidate. She/he can make four possible general conjectures about the candidate's type dependent strategies—two pooling strategies and two separating strategies. The two pooling strategies are (a) a candidate chooses MBA irrespective of the type (denoted henceforth as MBA–MBA)

Figure 7.7

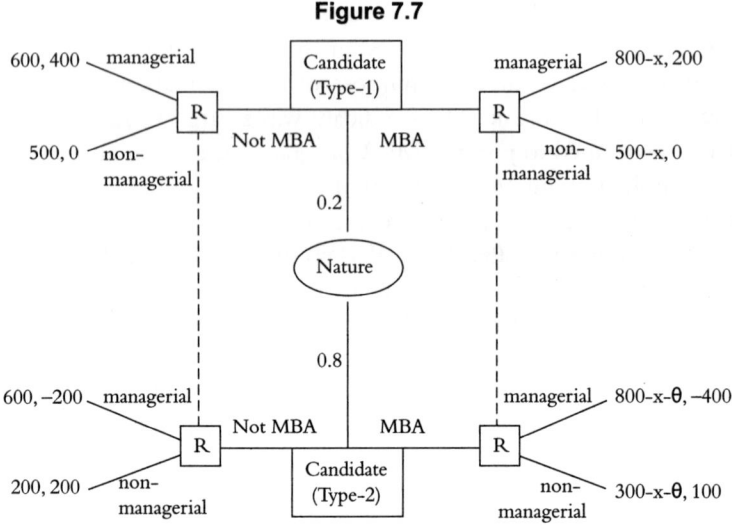

and (b) a candidate chooses not to go for MBA irrespective of the type (denoted henceforth as Not MBA–Not MBA). The two separating strategies are (a) a candidate chooses MBA if he is type-1 and chooses not to go for MBA if he is type-2 (denoted henceforth as MBA–Not MBA) and (b) a candidate chooses not to go for MBA if he/she is type-1 and chooses to go for MBA if he/she is type-2 (denoted henceforth as Not MBA–MBA). The recruiter observed that the candidate is an MBA. Hence, she/he can rule out the pooling strategy Not MBA–Not MBA. Her/his best response to MBA–MBA is to offer non-managerial role. Her/his expected payoff from offering the managerial role is −280 and that from offering a non-managerial role is 80. The recruiter's best response to the separating strategy MBA-Not MBA is to offer managerial role only upon observation of MBA, and to offer non-managerial role in the absence of MBA. Given her/his conjecture that only type-1 candidate can have MBA, being consistent to her/his conjecture she/he must offer managerial role only if the candidate is an MBA. The recruiter's best response to the separating strategy Not MBA-MBA is to offer managerial role only in the absence of MBA. Given her conjecture that only type-2 candidate can have MBA, she/he must offer managerial role only if the candidate does not have an MBA.

Putting himself in the shoes of the recruiter the candidate can anticipate the recruiter's best responses. Since the recruiter's best response to MBA–MBA is to offer a non-managerial role, if the candidate is type-1 he gains by choosing Not MBA instead of choosing MBA. By choosing MBA, he gets $(500 - x)$ and by choosing No MBA he gets 500. Basically, it does not make sense to spend effort on obtaining MBA. Therefore, the recruiter's conjecture that a candidate chooses MBA irrespective of the type is not rational. She/he should reject that conjecture. When the recruiter conjectures that only type-2 candidate goes for MBA, her best response is to offer managerial role only in the absence of MBA. If that is the case, the candidate gains by choosing Not MBA if he is type-2. The conjecture that only a type-2 candidate goes for MBA is also irrational and should be ruled out. That leaves us with the only other conjecture that a candidate chooses MBA only if he is type-1. Since the recruiter's best response to MBA–Not MBA is to offer managerial role if the candidate is type-1 and non-managerial role if he is type-2, the candidate, if he is type-1, does not gain by deviation if $x < 300$. If the candidate is type-1 and if he chose Not MBA instead of MBA, his payoff would have reduced from $(800 - x)$ to 500. If the candidate is type-2, he does not gain by deviating from Not MBA to MBA if $(x + \theta) > 600$. If the candidate is type-2 and if he chose MBA instead of Not MBA, his payoff would have reduced from 200 to $(800 - x - \theta)$. So, the conjecture that a candidate chooses MBA only if he is type-1 is rational if $(x + \theta)$ is larger than 600 and x is smaller than 300. Given her/his belief, the recruiter rationally conjectures that the candidate is type-1 if he has the MBA and is type-2 if he does not have the MBA, provided $(x + \theta)$ is larger than 600 and x is smaller than 300. She/he offers the managerial role to the candidate as he/she has the MBA. If he didn't have the MBA, the recruiter would have offered a non-managerial role to the candidate.

We need one final check to establish the signalling role of MBA. What if the candidate didn't have the MBA? If that was the case, the recruiter should have considered the conjecture that a candidate chooses not to go for MBA irrespective of the type indicated by the pooling strategy Not MBA-Not MBA. Given the pooling strategy Not MBA-Not MBA, the best response of the recruiter is to offer a non-managerial role. Her/his expected payoff from offering the managerial role is −80 and that from offering a non-managerial role is 160. If the candidate is type-1, he will gain by deviating from Not MBA to MBA provided $(800 - x) >$ 500. So, the recruiter's conjecture that a candidate chooses not to go for MBA irrespective of the type is irrational if $x < 300$.

We can now conclude that the candidate will choose MBA only if he is type-1 and he will be offered a managerial role if he has the MBA, provided the effort cost of obtaining an MBA degree is less than 300 for type-1 and more than 600 for type-2. These critical values of effort cost were derived from the payoffs taken in this example. In general, if the proportion of efficient people is low in the population of graduates and if the effort cost of obtaining MBA is much higher for the not so efficient ones vis-à-vis that for the efficient ones, MBA acts as a noiseless signal in the job market. In order to ensure that the job market signalling remains noiseless, the business schools need to make sure that the effort cost of obtaining an MBA remains critically high for the not so efficient people.

If the MBAs didn't get a premium, the type-1 graduates would have not spent the effort and would have not incurred the opportunity cost of not working for two years in their prime. Without the MBA premium, the business schools would have not existed. The recruiters would have incurred a huge cost of screening to get such a noiseless signal in the absence of the business schools. This is why the recruiters don't mind paying the MBAs a premium.

There are many possible applications of signalling games. Service providing firms get certified by neutral agencies to signal their quality to their customers. The idea behind such certification is not different from the way MBA degree is used to signal efficiency in the job market. The way business schools keep the effort cost high for the type-2 individuals, the certifying agencies need to keep the cost of getting certification high for firms that are less quality conscious.

Debt-to-equity ratio acts as a signal of insiders' confidence. A high debt-to-equity ratio on one hand indicates that the firm is not able to raise enough capital from the equity market, which in turn might indicate that the firm is not performing well. But a high debt-to-equity ratio also signifies that the management is confident about future performance of the firm. It is also possible that the insiders are not at all confident about the future firm performance but they increased the debt to bluff the investors. In that case the firm's management is taking high risk and the firm might burst. The managers should be liable and punished for knowingly taking such undue risk in order to avoid such behaviour. A high debt-to-equity ratio is a very noisy signal of firm performance.

Auctions as Games of Incomplete Information

Sealed-bid auctions are simultaneous move games of incomplete information. In a sealed-bid auction the bidders submit their bids in sealed envelopes without knowing what the bids of the other bidders are. However, it is possible to have an idea how much a bidder will bid for a particular object if we know how much that bidder values the object. Sealed-bid auctions are games of incomplete information because we don't even know what the valuations of the object to other bidders are. In game theoretic terms, the bidders don't know the types of other bidders. Nevertheless, as is done in games of incomplete information, we can place our bids based on our beliefs about the valuations of other bidders.

Auction is an age-old mechanism of exchanging objects and rights. There is recorded history of slave auction as early as in 500 BC. In ancient Greece, tax collection rights used to be auctioned off. The revenue generated from the auction used to go to the emperor's exchequer and the tax revenue to the collector. The bidders used to bid on basis of their expectation about tax revenue. The highest bidder used to get the right to collect taxes. It is rational to assume that the highest bidder used to have the highest expectation. But these auctions were not sealed-bid auctions, but were open auctions. The bidders used to call and outbid each other. Later, in Roman Empire, estates were liquidated using ascending order open auctions. In fact, the word auction comes from the Latin word "auctus", which means increasing or ascending. In present time, we refer to such ascending order open auctions as "English auction" because the famous English auction houses like Christie's and Sotheby's used the same mechanism to auction art objects and collectables.

There are various contemporary uses of different kinds of auctions. Most well-known are the spectrum rights auctions. Drilling rights in oilfields are also auctioned. Innovators sell the licenses to use new technologies. Commodities are sold through auction mechanisms. Government securities are auctioned by the treasury. When multiple raider firms bid to take over a target company, the situation becomes identical to an auction. Procurement auction, also known as reverse auction, is used by government departments as well as private companies to outsource jobs. One example of such auction is contracting of government-funded projects like highways construction. In case of reverse auctions, the party bidding the lowest price wins the contract.

Auctions may be either sealed-bid or open bid. Sealed-bid auctions are essentially simultaneous move games of incomplete information. There are two predominant variations that are used in sealed-bid auctions—first price auction and Vickery auction. In both forms the highest bidder wins. In case of reverse auctions the lowest bidder wins. In first price sealed-bid auction, the highest bidder pays his or her bid. In the reverse auction equivalent of a first price auction, the lowest bidder wins the contract and is paid his or her bid. Vickery auction was designed by William Spencer Vickery, the famous Canadian economist who won the Nobel Prize in 1996. Vickery auction is also known as the second price auction because in this variant of sealed-bid auction the highest bidder wins, but pays only the second highest bid. In the reverse auction equivalent of a Vickery auction, the lowest bidder wins but is paid the second lowest bid. The idea behind this apparently strange rule is to ensure that the bidders bid at their valuation (or cost, in case of reverse auction). In first price auction, bidders bid below their valuation. In this section we will analyse each of these forms of sealed-bid auctions and find out the rationales for such bidding behaviour using our knowledge of games played under incomplete information.

There are two dominant variants of open auctions—the English auction and the Dutch auction. In an English auction the bidders start with a low bid and gradually increase the bids in an endeavour to outbid the other bidders. Apparently, it looks like a sequential move game of complete information. But it is not. Indeed, the bids are placed sequentially in time, but the bidders don't take turns to bid. After a bidder calls, any other bidder may bid to outbid the ongoing highest bid. Even in an English auction, the bidders don't know each other's valuations for the object. All you know is the ongoing highest bid. In an English auction, the bidders outbid by bidding an amount just above the ongoing highest bid. When the ongoing highest bid exceeds the valuation of a bidder, he/she drops out. So, in an English auction the bidder with the highest valuation wins, but pays just above the valuation of the bidder with the second highest valuation. Ignoring the small difference we can say that in an English auction the winner pays the valuation of the bidder with the second highest valuation. That would be the case in a Vickery auction too, if it is true that the bidders bid just their valuations in a Vickery auction. In that sense the English auction is equivalent to a Vickery auction.

In a Dutch auction, the auctioneer asks for a price. If there is no taker at that price, the auctioneer gradually reduces the ask price and keeps on reducing gradually till a buyer agrees to the ongoing price. This form of auction is equivalent to a first price sealed-bid auction. In a first price sealed-bid auction, the bidders don't bid above their valuation. But if they bid at their valuation, they are not left with any surplus. A bidder's surplus is the difference between his or her valuation and the price at which he/she acquires the object. If they bid below their valuation they are left with surpluses. For bids below the valuation, lower the bid is, higher is the surplus, provided the bidder wins. But with low bids the chance of winning also gets reduced. So, it is a trade-off between surplus and chance of winning—a situation resembling a decision to raise or not in a game of poker. A bidder participating in a Dutch auction won't agree at a price higher than his or her valuation. But when the price is below his or her valuation, the bidder would face the dilemma whether to call or not. As the price reduces, the bidder's potential surplus increases. But the surplus won't accrue to the bidder if another bidder calls before him or her. The risk of not getting the object increases as the price falls. As in the case of a first price sealed-bid auction, it is a trade-off between surplus and chance of winning when the price goes below the bidder's valuation. In that sense, a Dutch auction is equivalent to a first price sealed-bid auction.

In the next two sections we will analyse the Vickery auction and the first price sealed-bid auction as simultaneous move games of incomplete information. We will consider the essential aspects of the two variants of sealed-bid auctions.

First Price Auction—An Example

Suppose there are only two bidders, A and B, in a first price sealed-bid auction. The highest bidder will win and will pay his bid. In case of a tie the auction is cancelled. Suppose, the valuation of A is 18 and that of B is V_B. Bidder A does not know V_B, but believes that V_B can be any integer between 10 and 30 with each value equally likely. B knows V_B but does not know that A's valuation is 18 and believes that A's valuation is an integer between 10 and 30 with each value equally likely. These numbers are taken just for the purpose of constructing an example. Let us find out what should be the best bid for A.

A should not bid above 18 under any circumstance as his surplus will be negative for all bids above 18. If he bids 18, his surplus will be zero. Only for bids below 18, A earns positive surplus if he wins. So, A should bid below his valuation. However, bidding 10 is ruled out. If A bids 10 he does not stand a chance of winning. If he bids x, which is less than 18 but more than 10, and wins, he earns a surplus of $(18 - x)$. Bidding x, A wins if and only if B bids between 10 and x. There are $(x - 10)$ integers between 10 and x, excluding x. But B can bid any of the 20 integer values between 10 and 29, including the boundary values. So, if A bids x, there is a $(x - 10)$ in 20 chance that A will win, that is, if A bids x he wins with probability $(x - 10)/20$. For example, if A bids 15, there is a 5 in 20, that is, 1 in 4 chance of winning for him. So, A's expected payoff from bidding x is $[(18 - x) \times (x - 10)/20]$. This expected payoff is maximum at $x = 14$. Hence, the best bid for A is 14, if he is risk-neutral. If A is risk-averse, he will bid more than 14 but less than 18. B can do a similar exercise and find what will be his best bid given his valuation.

If there were three bidders, A, B and C, A would need to bid more to maximize his expected payoff. In the presence of two other bidders, A wins if his bid is more than those of both B and C. If A bids x, his bid is more than that of B with probability $(x - 10)/20$, and is more than that of C with probability $(x - 10)/20$. So, by bidding x he wins with probability $[(x - 10)/20]^2$. So his expected payoff from bidding x is $(18 - x) \times [(x - 10)/20]^2$. This expected payoff is maximum when $x = 15$. The best bid for A in the presence of two other bidders is 15, which is more than his best bid in the presence of only one other bidder. As the number of bidders increases, the optimal bid increases.

Using this example we arrived at some observations that hold in general, and could be proven using a complete mathematical characterization of first price sealed-bid auctions. These observations are: (a) Bidders bid less than their valuation, (b) risk-averse bidders will bid more than risk-neutral ones and (c) the optimal bid for each bidder increases if more bidders are invited. In case of a procurement auction where the bidder who bids the lowest price gets the contract, the bidders will bid more than their cost, but the optimal bid will reduce with increase in the number of bidders. In such reverse auctions, the risk-averse bidders will bid less than the risk neutral ones in order to maximize the chance of winning, or rather to reduce the risk of not getting the contract.

Vickery Auction

In a Vickery auction, the bids and valuations of only the two bidders with the highest valuations matter, even if there are a large number of bidders. The highest bidder will win, but will pay only the second highest bid. In case of a tie the auction is cancelled. Let A and B be the two highest bidders. As was the case in the example that we constructed to analyse the first price auction, suppose the valuation of A is 18 and that of B is V_B. The bidders don't know each other's valuations. Each bidder believes that the valuation of the other bidder is any integer between 10 and 30 with each value equally likely.

The bid of B can be any integer between 10 and 30. If B's bid is more than 18, A is best off bidding 18, which is his valuation. If he bids above the bid of B, he wins the auction and pays B's bid. Doing so, he will earn a negative surplus as B's bid is more than 18. By bidding 18, that is, bidding his valuation, A loses the auction, but avoids the negative surplus. If he bids anything below B's bid, he gets zero. So, zero is the best possible payoff that A can get when B bids above 18, and by bidding just his valuation A gets that best possible payoff. If B bids exactly 18, A's payoff is zero for any bid. By bidding above 18, A gets the object but his surplus is zero as he has to pay 18. By bidding 18 or less he doesn't get the object. So, when B bids 18, A gets the best possible payoff by bidding 18. Finally, let's consider the case when B bids less than 18. Suppose B's bid is y, which is less than 18. A gets the object and is left with a surplus of $(18 - y)$ when he bids anything larger than y. A doesn't get the object if he bids less than or equal to y. So, A gets the best possible payoff by bidding 18, which is his valuation, even when B bids less than 18. Under no circumstances A is better off bidding anything other than his valuation.

In a Vickery auction, the bidders are best off bidding their valuations. This result is independent of the number of bidders. To prove that A is best off bidding his valuation, we didn't use expected payoff. We considered a belief of A, but didn't use the probability distribution. So, this proof is independent of the bidder's risk preference too. Irrespective of whether A is risk-averse or risk-neutral, his best response was to bid his valuation. In case of a procurement auction where the lowest bidder wins the contract but is paid the second lowest bid, the bidders will bid at their costs.

Revenue Equivalence

We found that bidders bid their valuation in a Vickery auction, but bid below their valuation in a first price auction. However, in a Vickery auction the auctioneer gets only the second highest bid whereas in a first price auction the auctioneer gets the highest bid. In which case is the auctioneer's revenue more? In order to check, let's go back to our example. Suppose A is the highest bidder, that is, $V_B < 18$. In the Vickery auction, A will bid 18 and B will bid V_B. So A wins and pays V_B. Since V_B is between 10 and 18 with all integer values equally likely, the auctioneer should expect to get 14, which is the average. In the first price auction with only A and B, A's optimal bid is 14. Since V_B is less than A's valuation, B's optimal bid will be lower than 14. Hence, the auctioneer earns 14 in the first price sealed-bid auction too. The auctioneer's revenue, on an average, is the same from the first price auction and the Vickery auction if the bidders are risk-neutral. This also is a general result and can be proven using a complete mathematical construct of the two forms of auctions. But if the bidders are risk-averse, the auctioneer's revenue is higher from the first price auction. In our example, A's bid would have been more than 14 in the first price auction if he was risk-averse. But the bids would have remained same in the Vickery auction. Hence, the expected auction revenue would have remained unchanged for Vickery auction, but would have increased for first price auction had the bidders been risk-averse.

8

Smart Negotiations

Anegotiation desk is an ultimate place to apply game theory to out-think your opponent. We have come across various business applications of different tools of game theory in this book. A separate chapter has been dedicated to negotiations because it applies almost everything discussed in this book so far. This chapter will primarily deal with the science of negotiation, which is founded in the value-net framework and bargaining games. We will also discuss creative tactical moves in negotiating deals, which is part of the art of negotiation. But a good negotiator also needs to be a good communicator—both in verbal as well as non-verbal terms. Those aspects of the art of negotiation are beyond the scope of this chapter.

Value-net Model and Added Value

The value-net model developed in Brandenburger and Nalebuff (1997), identifies four primary players who play either direct or indirect roles in negotiations that a firm participates in. These four primary types of players are the competitors, the suppliers, the customers and the complementors. Complementors produce complements to the firm's products. The term does not exist in English lexicon. It was coined by Brandenberger and Nalebuff.

Figure 8.1: Value-net Model

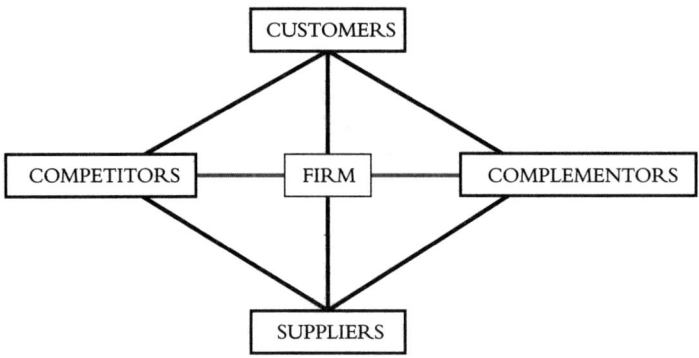

The firm purchases inputs from the suppliers and sells product to customers. Note that the vertical linkages of the model, as shown in Figure 8.1, are different buyer–seller relationships. The firm's competitors also purchase inputs from the suppliers and sell their products to the same set of customers. The customers buy complements from the complementors and the suppliers might supply certain common inputs (say, labour) to the complementors. All forms of buyer–seller negotiations can be analysed using bargaining games, which we will discuss in the next few sections of this chapter.

The horizontal linkages indicate lateral strategic relations, which could be leveraged while negotiating with another player who is vertically linked. For example, a firm and its competitor may form a seller's collusion to ensure higher prices, as discussed in Chapter 6. Forming collusion increases bargaining power. On the other hand, a buyer may make the firm bid against its competitor in order to gain bargaining power and, thus, get better price. Buyers may also collude. For example, the firm and its competitor may form a buyer's collusion to get better price from suppliers. The firm may also use the complementors to gain bargaining power while negotiating with customers. A tie up with complementors to sell products as bundles might be a useful strategy. There are four triangles in the value-net model. In any triangle there are three players at three vertices. For any bilateral negotiation, the third player on the remaining vertex of the triangle may be used to gain bargaining power or to reduce the bargaining power of the opponent. In the course of this chapter we will see that through a few case studies and examples.

Added Value

Value addition takes place in every node of a supply chain. A transaction happens only if all of the concerned players could improve payoff from the transaction as against their default payoff that is the one they receive when the transaction fails to take place. The total improvement in payoff is called added-value or gains from trade. The sharing of added-value requires negotiation. For example, suppose Soft Co. develops customized software for banking vertical. ABC Bank is a potential customer. If they get the software, they save US$1.5 million over a horizon of five years. Soft Co. also knows that the software will become obsolete after five years. Suppose the cost of developing the software is US$0.8 million to Soft Co. That means, in this case the value addition (surplus) is US$0.7 million. How much of the surplus is accrued to ABC Bank and how much to Soft Co. depends on the price. If the price is settled at P, the surplus to ABC Bank is US$(1.5 − P)$, and that to Soft Co. is US$(P − 0.8)$. The surplus to ABC Bank decreases and that to Soft Co. increases as P increases, making the relation between ABC Bank and Soft Co. a conflict relation. In order to resolve the conflict they need to negotiate sensibly.

Before we can proceed further, we need to have a clear understanding of bargaining games. In the next few sections we will discuss different forms of bargaining games. These games are sequential move games. We will discuss bargaining games under complete information as well as under incomplete information. For the first time in this book we will discuss some experimental results to reinforce the theory.

The Ultimatum Game

Let us begin with a very simple form of bargaining game to develop the paradigm in which we will discuss negotiations. Suppose ₹1,000 is to be distributed between two players—A and B. There are ten ₹100 notes and the money must be distributed using those very notes. This effectively means that the sum of ₹1,000 must be divided between A and B in multiples of ₹100. The setup of the ultimatum game is as follows. Player A is first given all the ten ₹100 notes. She/he must propose a division of those notes between her/him and player B. After A proposes a division, B can either accept that division or reject it. If B accepts A's proposal, then the money is divided as per the proposal. If B rejects, no one gets any money.

The game seems to be favourable towards the first mover. In fact, there exists a first mover's advantage in this game. The extensive form representation of the game is given in Figure 8.2.

The payoffs in Figure 8.2 are given as number of notes. A can offer any number of notes between 0 and 10. Instead of drawing 11 branches at the decision node of A, we represent the options using an arc. x is the number of notes actually offered by A, which is between 0 and 10. If B rejects, no one gets anything. If B accepts, A keeps $(10 - x)$ notes and B gets x notes. We can solve the game applying the logic of backward induction that we learnt in Chapter 2. A can foresee that being 'rational' B will accept any x as long as $x \geq 1$. Being self-interested, A wants to maximize her/his share of the pie. So, A will offer just one note to B. Since getting ₹100 is better than getting nothing, B will accept that. If this

Figure 8.2

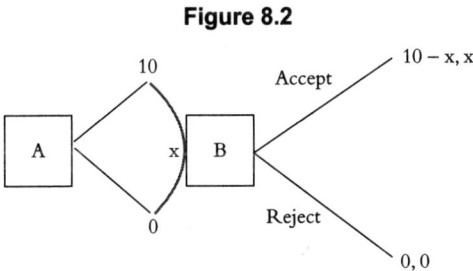

ultimatum game is played between two 'rational', 'self-interested' and 'emotionless' individuals, the first mover will offer one ₹100 note and keep nine, and the late mover will be forced to accept that 'unfair' distribution.

Experimental Findings

The ultimatum game was first tested experimentally by Güth et al. (1982). In last three decades the game, popularly known as 'divide the dollar' game (since the experimental game asked two players to divide ten US$1 bills amongst them), has been experimentally tested by different researchers on a very wide spectrum of samples from the human population. Two most prominent results are:

1. Majority of participants who played the role of 'proposer' (the role played by A in our example) offered fair distributions such as equal or close to equal shares.
2. When the 'proposer' offered less than 30 per cent share to the 'responder' (the role played by B in our example), an overwhelming majority of 'responders' rejected the offer.

The experimental findings were different from the outcome predicted by game theory. What is wrong with the theory? Two particular assumptions were questioned—'rationality' and 'self-interested behaviour'. These two assumptions are interrelated and generally referred to as 'individual rationality' in game theory. Do the experiments prove that the humans are not 'individually rational'? Or rather fairness is an innate human virtue? A slightly different experiment was used to find the truth.

Dictator Game

The dictator game is a variant of the ultimatum game that assigns more power to the proposer. In the dictator game the 'proposer' becomes the 'dictator' since the other player is made absolutely

Figure 8.3

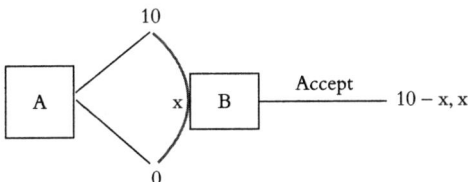

passive. Here, the 'dictator' distributes the money between him/her and the other player. Let's go back to our example where A moves first playing the role of the 'dictator'. B must accept whatever A proposes. This is not even a 'game' as we understood games in this book. It's a decision by A without any reaction from B. The extensive form representation of the decision is given in Figure 8.3.

Being self-interested and rational, and knowing that B is powerless and will accept any proposal however unfair that might be, A will offer nothing to B and take all the ten ₹100 notes.

Experimental findings of the dictator game were similar to the one predicted by theory and they throw some light on human fairness. A large majority of the participants, who played the role of the 'dictator', offered nothing or close to nothing to the powerless player. Indeed, human fairness is not rooted in altruism. Rather it is a derivative of what psychologists call 'negative reciprocity'.

Negative Reciprocity

If fairness was an innate human virtue, then even when the other player is powerless, majority of participants should have offered an equal split. Dictator game experiments proved that fairness is not really a human reflex. Rather it is derived from self-interest. Then, what could be the reason for 'fair' division in the ultimatum game? It is negative reciprocity at work. The idea is simple and intuitive. If the 'proposer' offers a small fraction of the pie, say 10 per cent share, the 'responder' might reject because she/he thinks the offer is unfair. That does not mean that the 'responder' is not individually rational. Here, the 'responder' is consciously allowing his/her emotion to play a role in decision-making. The 'responder' knows that there is a cost of rejection. She/he is willing to bear that cost in order to punish the unfair 'proposer'. Knowing that being vindictive the 'responder' might reject an unfair proposal, and not wanting to take unnecessary risk, the 'proposer' tends to make a fair offer.

Note that the above explanation does not negate 'individual rationality', nor does it reject the method of backward induction. Only it allows a space for human emotions, which are valuable to an individual's self. A hard-boiled economist will argue that the 'responder' rejects an unfair offer because the cost of rejection is less than the contingent value of his/her self-esteem that got hurt from the unfair treatment mated out by the 'proposer'.

Two separate research projects tested the ultimatum game with chimpanzees and human babies. Proctor et al. (2013) shows that chimpanzees behave in the same manner as humans do in the ultimatum game. The researchers tested chimpanzees and human children on a modified ultimatum game. One individual chose between two tokens that, with their partner's cooperation, could be exchanged for rewards. One token offered equal rewards to both players, whereas the other token favoured the chooser. Both apes and children responded like matured humans typically do. If their partner's cooperation was required, they split the rewards equally. However, with passive partners—a situation akin to the so-called dictator game—they preferred the selfish option. Thus, humans and chimpanzees show similar preferences regarding reward division, suggesting a long evolutionary history to the human sense of fairness. Another study by Kaiser et al. (2012) showed that chimpanzees do not reject unfair offers. The authors of the research paper published in *Biology Letters* suggest that human sense of fairness is a derived trait and that human fairness concerns evolved after the split between the lineages of humans and chimps. Taking these results from these two experiments with primates and human babies, one tends to argue that chimpanzees don't suffer from high self-esteem and, hence, they are not willing to pay a high price for it. That's why the chimp 'responders' didn't reject unfair offers. Despite that if chimp 'proposers' offer fair divisions, then it means that either they are unable to look forward and reason backward, or they are extremely risk-averse in terms of risk behaviour discussed in Chapter 7. Primates' inability of apply backward induction is not surprising as even humans find it very difficult to apply that logic in complex game situations unless they are trained to do so. But it might just be risk-aversion.

Ultimatum Game with High Stakes

If it is true that a 'responder' decides on rejection after weighing his/her contingent value of self-esteem and the cost of rejection, then it is intuitive to argue that the rejection rate should come down if the cost of rejection increases. If the players are asked to divide 10 wads of notes, each wad containing one lakh rupees, instead of being asked to divide ten ₹100 notes, then the 'responder' will find it difficult to reject even if she/he is offered the minimum, that is, one wad of notes. The cost of rejecting a 10th of ₹10 lakh is much larger than the cost of rejecting a 10th of ₹1,000. Now, if the 'proposer' can foresee that the 'responder' will find it difficult to reject even a grossly unfair offer, the 'proposer' should offer the minimum possible share. The ultimatum game with high stakes was first tested by Slonim and Roth (1998) and the results didn't support the theoretical prediction. The experiment couldn't really test how 'responders' react to low proportional offers in a high stake ultimatum game as only 4 out of 250 proposals made by the 'proposers' offered less than 20 per cent. Let us not forget that the stakes are very high for the 'proposers' too in the high stake ultimatum game. Hence, unwilling to take any chance, they offer a fair split. Only recently Andersen et al. (2011) showed, using a controlled experiment, how 'responders' react to low proportional offers in a high stake ultimatum game. This experiment was conducted in India and it showed that 'responders' accept lower proportional offers as the stakes are increased. The details of the experiment and results are summarized in Case Study 8.1.

Case Study 8.1: Individual Behaviour in Simple Bargaining Games—An Experiment in a High Stake Ultimatum Game Conducted in Villages of Meghalaya

A team of researchers led by Steffen Andersen from Copenhagen Business School conducted an experimental ultimatum game with varying stakes. The participants were villagers from eight different villages of Meghalaya, a state in the northeast of India. The participants were daily wage earners and had little wealth of their own. There were a total of 916 participants, who were divided into two equal groups—one group consisting of participants who were asked to play the role of 'proposers' and another group of those who were asked to play the role of 'responders'. Four hundred fifty eight pairs played the ultimatum game with four different orders of stakes—₹20, ₹200, ₹2,000 and ₹20,000. Considering ₹100 as the average daily wage earned by the participants, the stakes correspond to 1/5th of a day's work, 2 day's work, 20 day's work and 200 day's work, respectively. Since the participants normally don't work every day, or they don't find work every day, ₹20,000 correspond to more than a year's earning given that the average annual earning of the participants was ₹17,000. In that scale, ₹2,000 was close to their earning in one and half months, ₹200 was more than their earnings in four days and ₹20 was more than what they earned in three hours on an average. In order to test if there is any significant 'wealth effect', that is, if the behaviour depends on how rich the participant is, the researches gave an initial earning from unrelated task to 322 pairs of participants. Out of these 322 pairs, 173 pairs played the game with ₹20 at stake, 74 pairs with ₹200 at stake, 63 pairs with ₹2,000 at stake and 12 pairs with ₹20,000 at stake. Of the remaining 136 pairs with no initial earnings, 28 pairs played the game with ₹20 at stake, 50 pairs with ₹200 at stake, 46 pairs with ₹2,000 at stake and 12 pairs with ₹20,000 at stake. In order to make sure that the participants can apply backward induction, the researchers provided a cue to them. Basically, the logic of looking forward and reasoning backward was explained to the participants in the context of the game they were asked to play.

(Case Study contd.)

(Case Study contd.)

The results clearly showed that there is an inverse relation between the average share offered by the 'proposers' and the stake involved. While the average share offered was close to 25 per cent in the games with ₹20 at stake, it reduced to 17 per cent in games with ₹200 at stake, to 14 per cent in games with ₹2,000 at stake and to 12 per cent in games with ₹20,000 at stake. While the density of offers was highest in the 20–30 per cent range for the games with ₹20 and ₹200 at stake, it was highest in the less than 10 per cent range for the games with ₹2,000 and ₹20,000 at stakes. Indeed, the absolute offers increased with increase in stake. The median value of the actual offer increased from ₹5 in the ₹20 stake game to ₹30, ₹200 and ₹1,500 in the games with stakes of ₹200, ₹2,000 and ₹20,000, respectively. The researchers also found that there is not much effect of initial wealth that some pairs of participants were given.

The rejection rate of 'responders' decreased as stakes increased. While 36.32 per cent of offers were rejected in the game with ₹20 at stake, the rejection rate dropped to 4.17 per cent (only 1 out of 24 'responders' in that stake category) in the game with ₹20,000 at stake. The pattern of the decreasing rate of rejection with increasing stakes holds for both participants with initial wealth and those with no wealth. But for the participants with initial wealth, the rejection rate increased from 35 per cent in the game with ₹20 at stake to 47 per cent in the game with ₹200 at stake. The most important finding from the study was that the rejection rate decreased and approached zero as the amount of money that needs to be foregone due to the rejection increased.

Source: Andersen et al. (2011).

The finding of Andersen et al. (2011) shows that a responder's willingness to reject unfair offers decreases as the cost of rejection decreases. This in turn reinforces the hypothesis that the 'responder' weighs the cost of rejection against the contingent value of his/her self-esteem. The self-esteem of an individual is determined by the individual's social and economic status in the environment she/he lives in.

Slonim and Roth (1998) showed that 'proposers' insist on making fair offers even when the stakes are high. Andersen et al. (2011) show that when making proportionately small offers is induced in the 'proposers', the 'responders' do not reject unfair offers when the stakes are very high. Similar behaviour was observed among primates and human babies in Proctor et al. (2013) and Kaiser et al. (2012). Even though the 'responder' chimps didn't reject unfair offers, the 'proposer' chimps made fair ones. The sense of human fairness derived from negative reciprocity is possibly instinctive.

Based on the experimental results discussed in this section, we may conclude that humans tend to be self-interested and rational. Emotions might affect decision-making but humans can act rationally when the stakes are high. Also, humans lack the ability to apply the logic of backward induction in slightly complex scenarios. But if they are trained to look forward and reason backward they can do so. For example, a random person may not be able to see through the moves of the rival in a game of chess, but a trained chess player can do that for multiple numbers of moves and also make moves using backward induction.

We understand that businesses are high stake games and should be played rationally. Negotiators who make business deals are either trained to look forward and reason backward, or they learn to do so from experience. They are expected to make rational choices while making offers or accepting offers, keeping in mind that the opponent too is capable of doing the same. They need to be aware of their own 'best alternative to negotiated agreement' (BATNA) as well as the opponents' BATNA. With that understanding, in the remainder of this chapter, we will build our discussion on negotiations around a game theoretic approach towards bargaining games.

The Pirate Ship Problem

The pirate ship problem is a variant of the ultimatum game involving multiple players. This problem was constructed by English mathematician Ian Stewart in his now famous article "A Puzzle for the Pirates," which was published in *Scientific American* in 1999. The pirate ship problem exhibits the role of proper incentives in bilateral pacts in a multilateral negotiation setup. Indeed, it takes a rationalist approach and uses backward induction.

The narrative of the problem is as follows:

Black-hole Brandon is the captain of the pirate ship Macarena with four other pirates under his command. Second in command is Black Jack. Jack Sparrow and Jack Daniels are third and fourth in command, respectively, and Junior Black is the junior most. There are certain rules of the ship as follows.

- The captain always proposes a distribution of the loot. All pirates vote on the proposal, and if half the crew or more go 'aye', the loot is divided as proposed by the captain.
- If the captain fails to obtain support of at least half his crew (including him), he has to walk the plank, that is, he will be fed to the sharks.
- When the captain gets displaced (murdered, to be precise), the next in command becomes the captain. So, in order to become the captain, a pirate must kill all those above him in rank.

The pirates looted a ship and obtained 100 chests of gold. They have decided to disband after the loot is distributed. The captain will have to divide the loot among five of them. The smallest denominator should be a chest. The rules of the pirate ship hold good till the loot is divided and the group is disbanded.

The captain is self-interested and rational. He wants to take as many numbers of chests for himself as he can. Of course he doesn't want to get killed either. The captain does not have any special relation with any other pirate, nor does he favour anyone. He knows that all the pirates have functional, business-like relation with each other and no one is particularly friendly towards any other. He is aware that the pirates are all very treacherous, selfish, extremely intelligent and emotionless, exactly like him. He also knows that each pirate has the same understanding of the other members of the gang and that everyone in the gang is aware about others understanding of them. What is the maximum number of chests that the captain can keep for himself?

The captain should look forward and reason backward. If wrong decisions by Black-hole Brandon (rank 1), Black Jack (rank 2) and Jack Sparrow (rank 3) results in their deaths, Jack Daniels (rank 4) will become the captain. Since the rule of the pirate ship requires the captain to secure half of the votes, and since his own vote will be sufficient to ensure that his proposal stands, Jack Daniels will take it all leaving nothing for Junior Black (rank 5). So, Junior Black would not want it to reach the situation where Jack Daniels becomes the captain. He will be ready to go with Jack Sparrow (rank 3) to ensure that Jack Daniels (rank 4) does not become the captain. In order to incentivize Junior Black (rank 5) to come with him, Jack Sparrow needs to offer him only one chest of gold. So, in the situation wherein Black-hole Brandon (rank 1) and Black Jack (rank 2) are dead, Jack Sparrow can walk away with 99 chests with support from Junior. This is the situation that Jack Daniels hates. He is intelligent enough to understand that he does not stand a chance to become the captain as Junior will surely go with Jack Sparrow. Jack Daniels also understands that his interest lies in ensuring that Sparrow does not become the captain. Hence, in the situation where Black-hole Brandon (rank 1—present captain) is killed, Jack Daniels (rank 4) will be happy to support Black Jack (rank 2). Black Jack will get Jack Daniels' support by offering him just one chest. That means if Captain Black-hole Brandon is dead, Black Jack will walk away with 99 chests giving only one chest to Jack Daniels. This situation should be avoided by Jack Sparrow (rank 3) and Junior Black (rank 5) as they won't get anything if Captain Brandon is dead. Jack Sparrow should also be able to see that he does not stand any realistic chance of becoming the captain ever. With this foresight Captain Brandon can easily identify his allies—the ones who don't want him dead, that is, Jack Sparrow and Junior Black. Captain Brandon can ensure his life and still appropriate 98 chests by offering one chest each to Sparrow and Junior.

The argument given in the paragraph above is founded on the assumption that the pirates are 'individually rational' and they have enough cognitive ability to do the analysis. These assumptions are made explicit in the problem statement by means of the qualifiers that 'the pirates are all very treacherous, selfish, extremely intelligent and emotionless'. Another implicit assumption is that the cost of rejection is very high, that is even one chest of gold is large enough to incentivize the pirates to accept such a grossly unfair deal.

Alternate Offer Bargaining

In the ultimatum game, the 'responder' could only accept or reject the offer made by the 'proposer'. In context of a buyer–seller bargaining scenario, it is like a 'take it or leave it' offer. The BATNA, which we will refer to as default payoff from now on, for each player is zero in the ultimatum game. In a real-life business context the default payoffs may not be zero. Nevertheless, the ultimatum game models the situations of 'take it or leave it' offers. Alternate offer bargaining model captures the scenarios wherein the buyer (or seller, as the case may be) proposes a counter offer when they reject the original offer of the bargaining opponent. The game may go on for multiple rounds and both parties waste time till the deal is negotiated. Therefore, the cost of waiting should be factored into the game.

Let us begin with a two-stage alternate offer bargaining game. As in our setup of the ultimatum game, player A first gets a chance to divide ten ₹100 notes between her/him and player B. B can either accept A's offer or reject it. If B accepts the offer, the game ends and the ten notes are divided as per A's proposal. If B rejects A's offer, the game moves to the stage-2. In stage-2, which is at a later time, the players are left with only four notes. This reduction in the size of the stake is done to incorporate the effect of waiting cost. At the beginning of stage-2, B must make a counter offer to divide the remaining four notes, which A might either accept or reject. If A accepts B's offer, then the four notes get distributed as

Figure 8.4

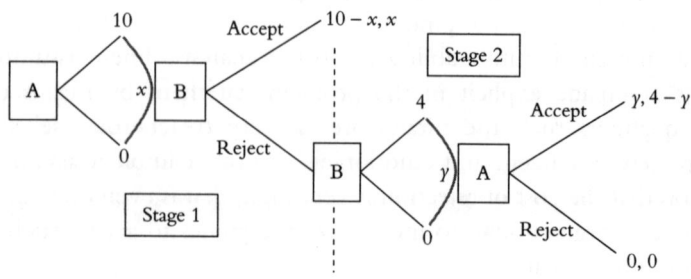

per B's proposal and the game ends. Even if A rejects B's proposal, the game ends and no one gets anything. The extensive form representation of the two-stage alternate offer bargaining game is given in Figure 8.4.

x is the number of notes offered to B by A in stage-1 and y is the number of notes offered to A by B in stage-2. Since A is the first mover, A's payoffs are written first. Here, the payoffs are the number of ₹100 notes.

Even before making the initial offer at the beginning of stage-1, A should be able to foresee what B can get by rejecting her/his offer in stage-1. If B rejects A's offer in stage-1, in stage-2 B can get a maximum of three notes. Since B knows that A is a rational and self-interested player, B can foresee that A will accept his/her offer if he/she offers at least one note to A in stage-2. So, if the game goes to stage-2 B can at best get three notes. With this foresight, A should offer four notes to B in stage-1, and being a rational and self-interested player B should accept that offer. This is the perfect solution of the two-stage alternate offer bargaining game played by rational and self-interested players who can look forward and reason backward.

The solution depends on the size of the stake, on waiting cost, on number of stages and on the default payoffs. We will understand the role of each factor using the Case Study 8.2. Case Study 8.2 is not a real case but a fictitious one, generally known as 'teaching cases' in business schools. It is impossible to find real cases of negotiations as the bargaining happens behind closed doors.

Case Study 8.2: A Teaching Case in Alternate Offer Bargaining—Acme Wagon Co. versus Selco Steel Inc.

Acme Wagon Co. manufactures goods train wagons. The company needs to buy one million tons of steel for their production and is bargaining with Selco Steel Inc. on the price of steel. Selco is the only domestic supplier of steel, and Acme is the largest domestic buyer of steel. In four weeks Acme will run out their stock of steel. If Acme fails to negotiate a price within four weeks, then they will have to import steel from abroad at a price of US$200 per ton.

For Selco the cost of production is US$80 per ton. There is a waiting cost of US$10 per ton per week for Selco. Till the price is negotiated, they have to hold the inventory. They cannot supply to other buyers as their production capacity is limited, and they have to hold the inventory for Acme. As a result they incur an opportunity cost.

In order to negotiate a price, the two firms play an alternate offer bargaining game. Since Acme will run out of their stock of steel in four weeks, they must settle the deal at least a week before that, since Selco takes one week time to deliver. The sequence of offers in the alternate offer bargaining game is as follows:

- Acme starts by offering a price in the beginning of the first week, which Selco may either accept or reject.
- If Selco rejects Acme's offered price, they can come back at the beginning of second week asking for an alternative price.
- Acme may either accept or reject Selco's asking price and may return at the beginning of the third week with yet another price offer.
- Selco may accept Acme's price offer in the third week or reject it. If they reject, they get a final chance to come back with a final price quotation at the beginning of the fourth week. Acme may either accept or reject Selco's asking price in the fourth week, but at the end of the fourth week they run out their stock.

(Case Study contd.)

(Case Study contd.)

> The firms must decide on their respective complete plans of action. Acme must decide what price to offer in the first week, what range of prices to accept in the second week if the price is not negotiated in the first week, what price to offer in the third week if the price is not settled even in the second week and what range of prices to accept in the fourth week if the deal is not done even in the third week. Likewise, Selco must decide on the ranges of acceptable prices in the first week and in the third week, if the game continues till the third week. They must also decide what prices to ask for in the second week and in the fourth week, subsequent to their rejection of Acme's offers in the first and third weeks, respectively.
>
> Source: Author.

Let us figure out the complete plans of actions for both the firms. That requires us to solve the entire four-stage alternate offer bargaining game. In order to avoid unnecessary complications in the bargaining process let us impose two rules. First, price (per ton of steel) should be quoted only in multiples of US$1. Second, as a rule of thumb to play the game, when a firm is indifferent between accepting and rejecting an offer, it must accept. Since the firms want to do business, the second condition is justified. Needless to say, both firms are aware of these rules as well as of the sequence in which the game is played. They are also aware about the payoffs and other information given in the case. Specifically, Acme is aware of Selco's cost of production and its cost of holding inventory. Selco is aware that Acme will run out of its stock of steel in four weeks. Both of them know the price of steel in the international market.

Now we can solve the game by the method of backward induction. The game tree is given in Figure 8.5. Payoffs are given in million dollars. At the beginning of week-1, Acme offers the price P_1. The prices are in dollars per ton. If Selco accepts the price P_1, the payoffs are X and Y. Here, we don't know the exact value added from the transaction, and hence we don't know how

Figure 8.5

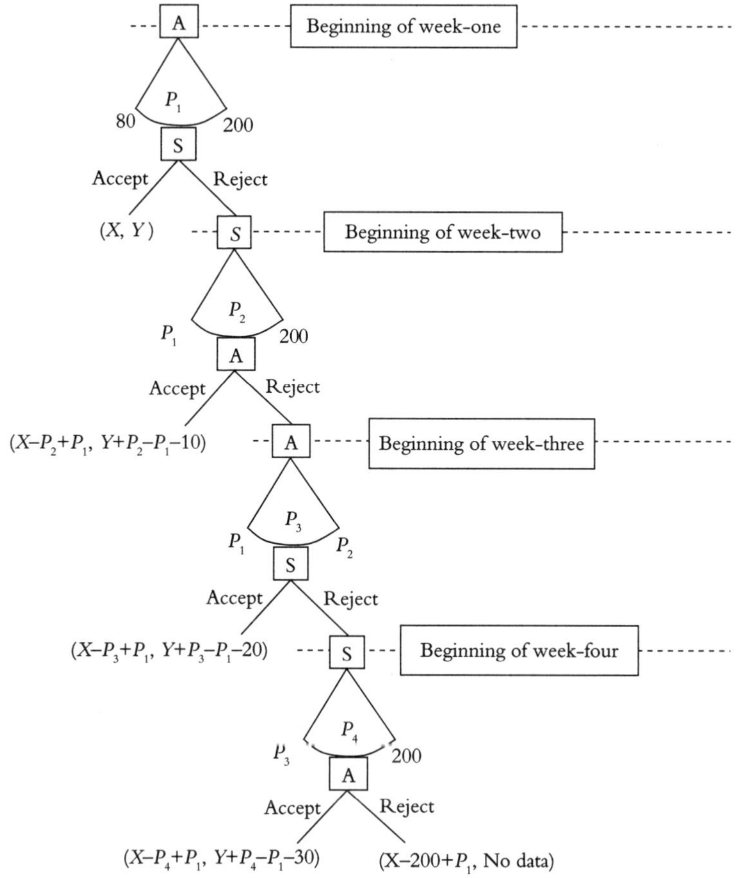

```
        --- A --------- | Beginning of week-one | ------------------
           /\
          /P₁\
      80 /    \ 200
        /  S   \
    Accept    Reject

   (X, Y)  ---| S |------ | Beginning of week-two | --------------

                /\
               /P₂\
          P₁ /    \ 200
            /  A   \
      Accept      Reject

(X-P₂+P₁, Y+P₂-P₁-10)  --- A ----- | Beginning of week-three | -----------

                          /\
                         /P₃\
                    P₁ /    \ P₂
                      /  S   \
                Accept      Reject

      (X-P₃+P₁, Y+P₃-P₁-20)  --- S --- | Beginning of week-four | --------

                                /\
                               /P₄\
                          P₃ /    \ 200
                            /  A   \
                      Accept      Reject

          (X-P₄+P₁, Y+P₄-P₁-30)      (X-200+P₁, No data)
```

much X is. Of course Selco's margin is US$$(P_1 - 80)$ per ton, and the quantum of the deal being a million ton, Y is US$$(P_1 - 80)$ million. But we can solve the game by benchmarking all payoffs against the week-1 payoffs of the respective firms, and the exact values of X and Y don't matter for our analysis. Note that P_1 cannot be less than US$80 as Selco's cost of production is US$80 per ton. Also, P_1 cannot be more than US$200 as Acme can import at that price. If Selco rejects Acme's offer and reverts in week-2 with an asking price P_2 dollars per ton, and if Acme accepts that asking price, the payoffs are $X - P_2 + P_1$ for Acme and $Y + P_2 - P_1 - 10$ for Selco. P_2 is more than P_1 and, hence, Acme's payoff reduced by $(P_2 - P_1)$ dollars per ton. Selco's payoff increased by the same amount, but it also lost US\$10 per ton due to waiting. Acme may reject Selco's asking price in week-2 and revert in week-3 with another offer of P_3 dollars per ton. If Selco accepts Acme's offer in week-3, Acme and Selco get payoffs of $X - P_3 + P_1$ and $Y + P_3 - P_1 - 10$, respectively. These payoffs are also in reference to week-1 payoffs. P_3 is less than or equal to P_2 but more than P_1. Since Selco already rejected P_1 in week-1, offering a price less than P_1 after two weeks is not a sensible offer, given that both players are individually rational and smart enough to use backward induction. So, if the deal gets negotiated in week-3, Acme's payoff reduces by $(P_3 - P_1)$ dollars per ton in comparison to its week-1payoff. Selco's payoff increases by the same amount, but it also loses US\$20 per ton due to waiting for two weeks. If Selco rejects Acme's week-3 offer and reverts in week-4 with an asking price P_4 dollars per ton, and if Acme accepts that asking price, then the payoffs are $X - P_4 + P_1$ for Acme and $Y + P_4 - P_1 - 30$ for Selco. P_4 is more than P_3 and, hence, is more than P_1. Therefore, Acme's payoff reduces by $(P_4 - P_1)$ dollars per ton. Selco's payoff increases by the same amount but it also loses US\$30 per ton due to waiting for three weeks. If Acme rejects Selco's week-4 offer, then they have to import steel at the price of US\$200 per ton. Hence, in comparison to week-1, Acme's payoff reduces by $(200 - P_1)$ dollars per ton. We have no information about Selco's default payoff here. But that does not matter in solving the game as Selco is not deciding in that final node of the game.

In week-4, Acme will accept any price less than or equal to US\$200 per ton. As the seller Selco wants the maximum possible price, it will ask for US\$200 in week-4. So, if the deal gets settled in week-4, the payoffs are $X - 200 + P_1$ and $Y + 200 - P_1 - 30$. Acme should be able to foresee this week-4 outcome while they make their offer in week-3. Using that foresight, Acme should realize that Selco will accept their offer of P_3 in week-3 if Selco's payoff from doing so is more than $Y + 200 - P_1 - 30$, that is, if $Y + P_3 - P_1 - 20 \geq Y + 200 - P_1 - 30$ or $P_3 \geq 190$. Being the buyer

Acme wants the least price and, hence, it should offer US$190 per ton in week-3. While asking for P_2 in week-2, Selco should foresee that if the deal is not negotiated in week-2 Acme will offer US$190 per ton in week-3. Selco should make sure that Acme gets at least as much in week-2 as they would get if the deal gets settled in week-3, that is, they should ask for P_2 such that $X - P_2 + P_1 \geq X - 190 + P_1$ or $P_2 \leq 190$. So, they will ask for US$190 in week-2 itself. Note that there is no difference between the week-3 price and week-2 price. The reason is that there is no cost of waiting for Acme. In week-1, Acme should see through the game and see that Selco will accept their offer in week-1 itself if their payoff is least as much as their week-2 payoff, that is, $Y \geq Y + 190 - P_1 - 10$ or $P_1 \geq 180$. So, in week-1 itself, Acme should offer US$180 and that should be accepted by Selco.

It is important to note that any alternate offer bargaining game gets over in the very first stage. If the players are rational, are aware of their own as well as the other's default payoff, and if both are smart enough to see through the entire game and are capable of making perfect decisions by reasoning backward in any node they move, then there is no reason for the game to continue for beyond stage-1. The perfect offer in the very first stage splits the pie, but how the relative shares of the pie accrue to the contending players depend on the following:

1. The *default payoffs*—Larger is the default payoff of a player, larger is his/her share of the pie. A larger default payoff increases the bargaining power.
2. The *waiting cost*—Higher is a player's waiting cost, lower is his/her share. When the bargaining opponents know that you have a high waiting cost, they also understand that you will be eager to settle the deal quickly. A high waiting cost reduces the bargaining power.

In the Acme versus Selco case, if the international price of steel was higher, then it would have increased the bargaining power of Selco and reduced that of Acme. For example, if the international price was US$250 per ton instead of US$200 per ton, it would have forced Acme to offer US$230 per ton in week-1. On the other hand, if international price was US$170 per ton, Acme would have not offered any more than US$150. In this case, Acme would have always offered US$20 less than the international price since they knew that Selco's waiting cost is US$10 per week. But if Selco's waiting cost was US$30 per week, in this four-stage alternate offer bargaining game itself Acme would have offered US$60 less than the international price. Acme would have reduced price by twice that of Selco's weekly waiting cost because potentially Acme could make offers twice in the span of four weeks—once in week-1 and again in week-3. If Acme had stock of steel for six weeks, and the alternate offer bargaining game continued for six weeks with Acme getting to make offers in the week-1, week-3 and week-5, then their offer price in week-1 would have been less than the international price by thrice that of Selco's weekly waiting cost, and Selco would have been forced to accept. Having stock for more weeks would have made Acme more patient in negotiation and would have increased its bargaining power. The learning from the alternative offer bargaining game can be summarized in the following phrase: *Your bargaining power depends on the depth of your pocket and on how patient you can be.*

Collective Wage Bargaining

Collective bargaining is a periodic exercise in which an employer and a group of employees negotiate and revise working conditions, including wages. In presence of an organized labour union, the bargaining takes place between the union and the management. In many developed countries, particularly in Europe, labour laws mandate the employer to voluntarily take part in the process of collective bargaining. In India the employers are not bound by legal requirements to sit on the bargaining desk, but the workers have legal rights to go for strike if their demands are not met. In order to avoid strikes, it is advisable that the employer voluntarily takes part in the process of collective bargaining. The entire gamut of collective bargaining is beyond the scope of this chapter. In this section, we will particularly focus on the issue of collective wage bargaining.

After all other factors of production, except labour, are paid for from the revenue of the firm, what is left is surplus. We can think of this surplus as a pie that is to be shared between the employer and employees or, in a rough sense, between the workers and the management. If the union demands a higher wage rate, and the management gives in to that demand, then basically the workers increase their share of the pie. If the management puts its foot down and declines the union's demand, and if the union successfully leads the workers to go for a strike, then the size of the pie gets reduced. Let us consider the context wherein the law, unlike in India, mandates an employer to sit on the bargaining desk with the union at a given interval, say, in an interval of six months. Suppose it is common knowledge that the firm will earn a surplus of S over these six months. In a mature market with very little demand uncertainty and stable prices, S can be estimated quite precisely. If the number of workers employed remains unchanged during this period, then the size of the wage bill depends on the wage rate. Suppose there are N workers and the wage rate is w per month. So, the total wage bill is $W = 6wN$. If w is increased, then W will also increase. If S is the surplus and W is the wage bill, then firm's profit is $(S - W)$. Here, S is the total pie, and W and $(S - W)$ are workers' and firm's shares of the pie. Hence, if the wage rate increases, the workers' share of the pie increases. The purpose of collective bargaining is to split this pie, and the collective wage bargaining game between the labour union and the management can be seen as an alternate offer bargaining game. Till the wage rate and, thus, the shares are negotiated, production is stalled due to strike. Strike reduces the size of the surplus, which is the pie here. The reduction in size of the pie does not only affect the firm, but also the workers, as they don't get wages during the period of strike.

With that understanding we are ready to develop a slightly abstract but sufficiently general and robust model of alternate offer collective wage bargaining between the firm's management and the labour union. Whatever be the value of surplus, we may normalize the size of the pie to one since the contenders bargain for shares of the pie. First, let us consider the alternate offer bargaining game under complete information. To see how the number of stages matter, we will first consider a two-stage game and later we will extend it to three-stage game. In order to check if there exists any first mover's advantage, we will check the outcomes in the two-stage and three-stage cases with the firm moving first, as well as with the union moving first.

Alternate Offer Wage Bargaining— Even Number of Stages

The firm (henceforth, denoted as F) and the labour union (henceforth, denoted as U) are bargaining over shares of the surplus. In stage-1, that is, at the beginning of period-1, F makes an offer asking U to take x fraction of the pie. U may either accept or reject that offer. If U accepts, then F retains $(1 - x)$ fraction and U gets x fraction. If U rejects, then there is no production in period-1 due to strike and, hence, the size of the pie reduces from 1 to $(1 - \phi)$, where ϕ is the fraction of pie lost due to strike for one period. If U rejects F's offer in period-1 and goes for strike, then in stage-2, that is, at the beginning of period-2 U must make a counter-proposal demanding y fraction of the remainder pie. If F accepts U's demand in stage-2, then F gets $(1 - y)$ fraction of $(1 - \phi)$, that is, $(1 - y)(1 - \phi)$, and U gets y fraction of $(1 - \phi)$, that is, $y(1 - \phi)$. If F rejects U's demand at the beginning of period-2, then the situation is termed as a dispute, and the dispute is referred to the labour commission or eventually the court. The exogenous legal settlement takes one more period, and hence production does not happen in period-2 too due to strike, resulting in further shrinking of the pie by ϕ. When the dispute is settled, at the beginning of period-3, $(1 - 2\phi)$ is left of the pie. Indeed there might be uncertainty regarding how the legal system will split the remainder of the pie between F and U. But the contenders may have a fair idea about how the legal system splits surplus. In order to avoid modelling the uncertainty, let us assume that the legal system splits the

Figure 8.6

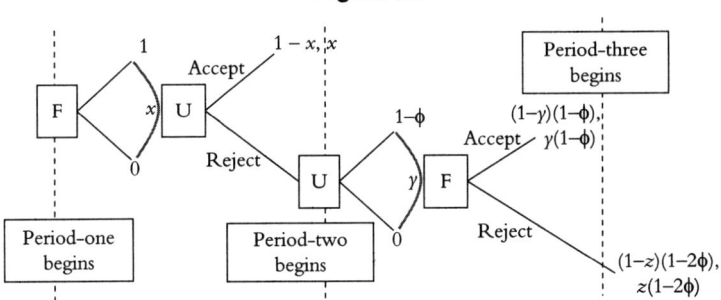

remainder of the pie in ratio $(1 - z)$: z between F and U. So, if the contenders fail to split the pie even in the stage-2 of the game, then at the beginning of period-3 F gets $(1 - z)(1 - 2\phi)$, and U gets $z(1 - 2\phi)$. The extensive form representation of the two-stage alternate offer collective wage bargaining game is given in Figure 8.6. As is the convention, payoff of F is given first, as F is the first mover in this game.

In period-2 F will give in to U's demand if $(1 - y)(1 - \phi) \geq (1 - z)(1 - 2\phi)$, that is, $y(1 - \phi) \leq z(1 - 2\phi) + \phi$. Anticipating that F will accept their demand if $y(1 - \phi) \leq z(1 - 2\phi) + \phi$, U should demand y such that $y(1 - \phi) = z(1 - 2\phi) + \phi$. Knowing that they will get $[z(1 - 2\phi) + \phi]$ in period-2, if the splitting is not negotiated in period-1, U will reject F's offer of x if $x < z(1 - 2\phi) + \phi$. Anticipating that their offer will be accepted by U if and only if $x \geq z(1 - 2\phi) + \phi$, F will offer just $[z(1 - 2\phi) + \phi]$ and retain $[1 - z(1 - 2\phi) - \phi]$, which is equal to $[(1 - z)(1 - 2\phi) + \phi]$. Both F and U get ϕ more than their respective default payoffs. When the contenders can see through the game and make perfect decisions using logic of backward induction, each of them get more than what they would have gotten had they failed to negotiate and the dispute was settled by the legal system. By avoiding strike they save a loss of 2ϕ, which gets equally split.

In order to see if there exists any first mover's advantage, we need to check if U's payoff increases when they first make the demand. Suppose U demands a share y in stage-1. If F rejects, then there is no production in period-1 and the size of the pie

Figure 8.7

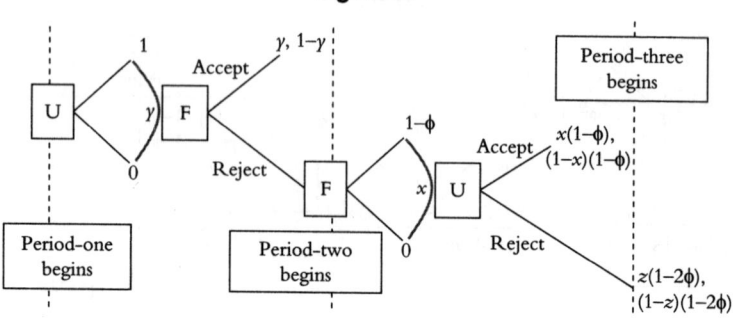

reduces to $(1 - \phi)$. In stage-2 F offers x fraction of $(1 - \phi)$ to U. If U rejects F's offer, then there is no production in period-2 too and the dispute is settled legally at the beginning of period-3. As before, the legal system splits $(1 - 2\phi)$ between U and F in the ratio $z: (1 - z)$. The extensive form representation of the game is given in Figure 8.7. Here, the payoff of U is given first as U is the first mover in this game.

In period-2 U will accept F's offer if $x(1- \phi) \geq z(1 - 2\phi)$. Anticipating that F offers x such that $x(1 - \phi) = z(1 - 2\phi)$. In that case, F gets $(1 - x)(1 - \phi) = [(1- z)(1 - 2\phi) + \phi]$. In stage-1 F should foresee that if the split is not negotiated in period-1, it will get $[(1 - z)(1 - 2\phi) + \phi]$ in period-2. Hence, it will give in to U's demand if and only if $(1- y) \geq [(1 - z)(1 - 2\phi) + \phi]$, that is, $y \leq 1- [(1 - z)(1 - 2\phi) + \phi]$ or $y \leq [z(1 - 2\phi) + \phi]$. Being rational U must demand just $[z(1 - 2\phi) + \phi]$. F will accept that and will retain $[(1 - z)(1 - 2\phi) + \phi]$.

Comparing the results when F moves first against when it moves late, we see that there is no effect of the sequence of moves. There is no first mover's advantage or late mover's advantage, if the alternate offer wage bargaining continues for two periods. The results are true for any alternate offer wage bargaining game with even number of periods. The threat of strike creates a potential loss in surplus. In our two-stage model the magnitude of that potential loss is 2ϕ, and the threat of strike gets the workers half of that potentially lost surplus.

Hence, if worker's share of surplus is less than half, they gain by threatening to strike. We need to check if the results found in the alternate offer wage bargaining game with even number of stages hold if there are odd (three or more) numbers of stages.

Alternate Offer Wage Bargaining— Odd Number of Stages

The game remains same as before, but continues for one more stage. Suppose F makes the first offer at the beginning of period-1 and gets to make another offer at the beginning of period-3, if the split does not get negotiated in the course of first two stages. The negotiation fails and the legal system intervenes to settle the dispute if U rejects F's offer even in period-3. As before, production stops due to strike till the split is negotiated, and ϕ fraction of the pie is lost in each period due to the strike. Let F offer shares x in period-1, and w in period-3, if negotiation is not settled till then. U demands y fraction in period-2. Legal system splits it in ratio $(1 - z)$: z between F and U. The game tree is given in Figure 8.8.

Like any alternate offer bargaining game played under complete information, this game too will be over in the stage-1 if the first mover makes the correct offer. We can find the optimal offer of F in stage-1 by backward induction. In stage-3, F will offer w such that $w(1 - 2\phi) = z(1 - 3\phi)$ and will retain $(1 - w)(1 - 2\phi) = [(1 - z)(1 - 3\phi) + \phi]$. Therefore, in stage-2, F will give in to U's demand if and only if y is such that $(1 - y)(1 - \phi) \geq [(1 - z)(1 - 3\phi) + \phi]$. Hence, in stage-2, U will demand y such that $(1 - y)(1 - \phi) = [(1 - z)(1 - 3\phi) + \phi]$ and, thus, get $y(1 - \phi) = z(1 - 3\phi) + \phi$. In stage-1 U will accept F's offer if and only if $x \geq z(1 - 3\phi) + \phi$, and anticipating that it will be accepted F will offer $x = z(1 - 3\phi) + \phi$ and retain $(1 - x) = [(1 - z)(1 - 3\phi) + 2\phi]$.

In order to check if there is a first mover's advantage, let U make the first move. U demands the share y in period-1, and v in period-3 if the split is not negotiated till then. F offers x fraction

Figure 8.8

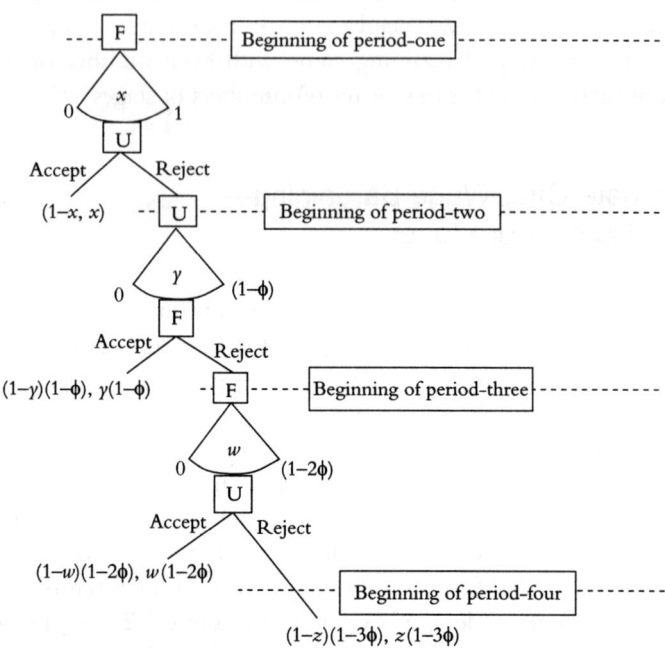

in period-2. Legal system splits it in ratio $(1 - z): z$ between F and U. The game tree is given in Figure 8.9.

In stage-3, F will accept v if and only if $(1 - v)(1 - 2\phi) \geq (1 - z)$ $(1 - 3\phi)$, that is, $v(1 - 2\phi) \leq z(1 - 3\phi) + \phi$. Hence, U will demand v such that they get $v(1 - 2\phi) = z(1 - 3\phi) + \phi$. In stage-2 itself U can foresee that if the game continues to stage-3 they can ensure $z(1 - 3\phi) + \phi$. So, in stage-2 U will accept F's offer if and only if $x(1 - \phi) \geq [z(1 - 3\phi) + \phi]$. So, in stage-2, F will offer x such that $x(1 - \phi) = [z(1 - 3\phi) + \phi]$ and will retain $(1 - x)(1 - \phi) = [(1 - z)$ $(1 - 3\phi) + \phi]$. Therefore, in stage-1 F will give in to U's demand if and only if $(1 - y) \geq [(1 - z)(1 - 3\phi) + \phi]$, and anticipating that it will be accepted that U will demand $y = z(1 - 3\phi) + 2\phi$ and F will retain $(1 - y) = [(1 - z)(1 - 3\phi) + \phi]$.

Like any finite horizon alternate offer bargaining game under complete information, the three-stage alternate offer wage bargaining game gets over in the very first stage if the players can

Figure 8.9

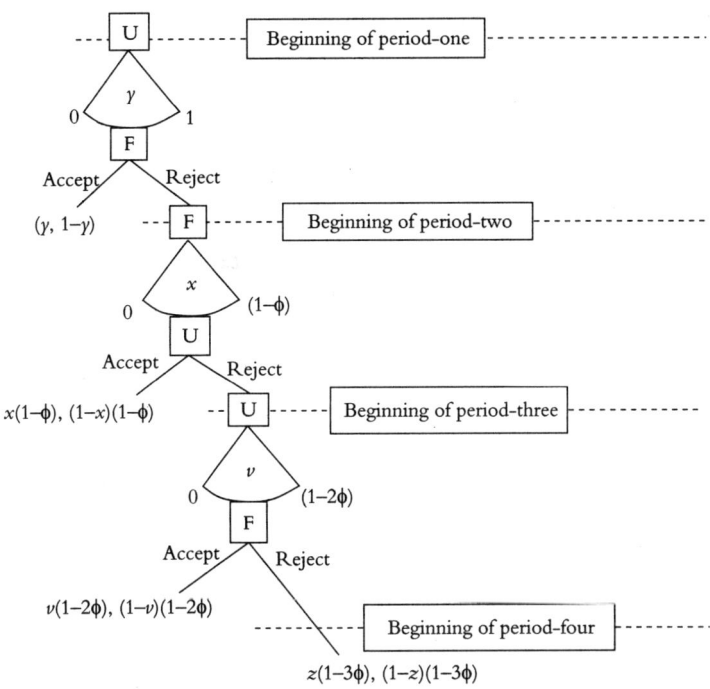

apply backward induction and play rationally. But if F moves first, it gets $[(1 - z)(1 - 3\phi) + 2\phi]$ and by moving later gets $[(1 - z)(1 - 3\phi) + \phi]$. Clearly, there is a first mover's advantage. Likewise for U, if they move first they get $[z(1 - 3\phi) + 2\phi]$ whereas by moving later they get $[z(1 - 3\phi) + \phi]$.

In this model of alternate offer wage bargaining under complete information, we assumed that a player accepts an offer or gives in to a demand if their payoff from accepting is same as that from rejection, that is, they accept when they are indifferent. As a consequence of that the 'responder' in a stage game decides to accept if they get same as what they would get if the game continues till the next stage. In effect, the potential loss of surplus due to strike in a period is appropriated by the 'proposer' of that stage game. If there are even numbers of periods, both the firm and the union get to propose (offer or demand) the same number of times.

Hence, the game becomes evenly balanced. But if there are odd numbers of stages, the first mover gets to propose one more time than the late mover. That is, the first mover gets an advantage in this alternate offer wage bargaining game if there are odd number of periods.

The assumption that the players accept when they are indifferent is not a restrictive assumption here. Indeed, they might reject when they are indifferent, and in that case their bargaining opponents must incentivize them to accept. If the players are rational, an incentive of small magnitude should be sufficient. The treatment of the game and the fundamental results do not get affected by incentives of small magnitude.

Why Strikes Happen?

In a world of complete information with rational players, any alternate offer bargaining game including wage bargaining gets over in the first stage. Having perfect foresight and the ability to reason backward allows the players to make the perfect offer (or demand) in the very first stage, which gets accepted by the other party due to their perfect foresight and the ability to reason backward. So, what could be the reasons behind worker strikes that happen at times? The following could be the plausible reasons:

- Negotiators are not rational.
- Negotiators don't have perfect foresight.
- Negotiators are unable to reason backward.
- Negotiators are politically motivated.
- Information is incomplete.

Many behavioural theories presume that players in bargaining games are not rational. Even if the players are not perfectly rational, they are bounded rational and undergo adaptive learning. Union leaders who engage in collective bargaining are seasoned negotiators and no rookie trade-unionist. It is rather irrational to assume that the negotiators are irrational. Moreover, negotiators from both the union side as well as from the firm side undergo training in negotiation skills. It is unlikely that they do not have perfect foresight or cannot reason backward.

It is indeed the case (particularly in developing nations, including India) that unions are controlled by political parties. In such cases the expected political gains or losses of mother party (or party leader, who may as well be the union leader) influences the decisions of the negotiators from the union side. In order to model such a game we need to extend the game and modify pay-offs incorporating the political gains and losses.

That leaves us with the last possibility that the negotiation takes place without complete information. The major source of information asymmetry is the size of the pie. The management of the firm knows the profit figures. But, the workers' union may not know it. Unless the financial statements of the firm are reported to the workers' union regularly, there is no reason for them to trust a financial statement which is presented to them after a dispute has brewed up. If there is trust between the firm's management and the workers, then strikes are unlikely to happen.

So far we have normalized the size of the pie to 1. We could do that because we assumed that the size of the pie is known to the contending parties and that they negotiate on relative shares. But if the workers' union, which is one of the contending parties, does not know the exact size of the pie, we cannot normalize it to 1. Let the actual size of the surplus be S^*, which is known to the firm (F). The union (U) believes that S is between S_{min} and S_{max} with all values in the domain equally likely. In an extreme case it is possible that $S_{min} > S^*$. In that case there exists no room for negotiation and strike is inevitable. As before, F and U bargain over the share of the surplus, but the contenders have different perceptions about the magnitude of the surplus. Whenever U makes decision, they make it on the basis of their belief that S^* is between S_{min} and S_{max}, and that itself is a good reason for negotiation failure. U might be risk-averse or risk-neutral. As we did for games of incomplete information in Chapter 7, let us assume that U is risk-neutral. So, U will negotiate on the basis of the expected size of the surplus as per their belief, that is, $[(S_{min} + S_{max})/2]$.

Even if actual S^* is between S_{min} and S_{max}, it is possible that it is less than $[(S_{min} + S_{max})/2]$. In that case, negotiation might fail and strike might happen.

Wage Bargaining Fundamentals

From our discussion in this section we arrive at the following fundamentals of collective wage bargaining:

1. While engaging in collective bargaining, a firm must set up the game such that it makes the first offer and the bargaining game continues for odd number of stages.
2. In order to make the first move the firm must offer a fair raise before the union comes up with any demand.
3. The firm must make sure that it gets to make a final offer before the legal system intervenes to resolve dispute.
4. It is in the interest of the firms to ensure that the union leaders are trained in the science of negotiation.
5. The firm must share all information on financials of the firm with the union, that is, it is advisable to maintain an open book policy. If the union does not know the size of the pie they will overestimate, and in that case their demand will also escalate.

Tactical Issues in Negotiating

Now let us shift our focus from the science of negotiating to the art of negotiating. The purpose of these tactical moves is either to increase your bargaining power or to decrease the bargaining power of your opponent. Often these tactical moves leverage on the existence of a third player. In terms of the value-net model of Bandenburger and Nalebuff, any of the two players involved in negotiating a deal may use a third player who, along with the contending parties, makes one of the triangles of the model. Let us explore the possibilities with some examples.

Having More than One Supplier

Having one supplier bestows monopoly bargaining power to the supplier. Particularly when the seller has other buyers but there is no alternative option for the buyer, it becomes an ultimatum game. The seller makes a 'take it or leave it' price offer, and the buyer is forced to accept it. Let us refer back to Case Study 3.1, Battleground Iberia, in Chapter 3. Suppose Iberia had only one supplier, Airbus. Let the value of each A340 aircraft be V to Iberia, and let C be the cost of manufacturing an aircraft to Airbus. Let the default payoff of Airbus and that of Iberia be D and d, respectively. If they fail to negotiate, Airbus is able to sell to other airlines. So, D is the minimum profit that Airbus can make from selling one aircraft. For Iberia, suppose the only options other than A340 are second-hand aircrafts. Therefore, $d = $ (Value of a used aircraft − Price of a used aircraft). The game is shown in Figure 8.10.

Iberia will take it if $V - P \geq d$, that is, if $P \leq V - d$. Since d is likely to be very small, the price that they are forced to pay becomes close to the valuation of A340. Indeed, they did much better by bringing in another supplier, Boeing. We saw that in Case Study 3.1. The strategic scenario, in terms of the value-net model, is shown in Figure 8.11.

With two suppliers, Iberia got the bargaining power. As a buyer they made the two competing giants play a Prisoner's Dilemma. This would have not been possible if Boeing and Airbus could collude. But they could not because Iberia was a crucial buyer for both of them. There were 200 odd buyers and only two sellers in the market. Indeed, the market power was much more for the

Figure 8.10

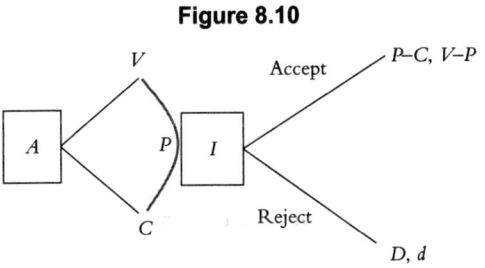

Figure 8.11

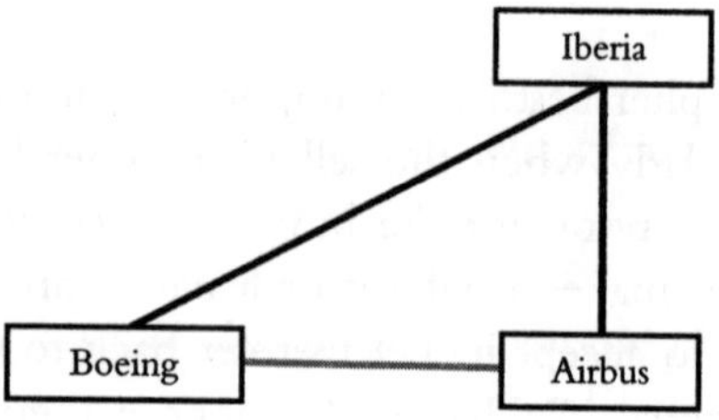

sellers. But this case shows that bargaining power does not always depend on market power.

Having More than One Buyer

Having more than one buyer increases the bargaining power of the seller. You have to learn it from football superstar Cristiano Ronaldo. In 2012–13 he was paid €10 million per year by his employer Real Madrid football club. In September 2013 he signed a new contract with Real Madrid with a 70 per cent raise in his annual salary. How did he get that? Indeed, he performed. But he also said that he was unhappy at Real Madrid, and would be happy to return to his former employer Manchester United football club. Manchester United offered him an annual salary of £14 million, which was equivalent to €16.67 million (at the then exchange rate). For Real Madrid, Ronaldo was almost indispensable. But before Manchester United came into the picture they were not ready to give such a huge raise to Ronaldo. Once Manchester United offered him €16.67 million per year, Real Madrid renewed Ronaldo's contract at an annual salary of €17 million. After signing the contract Ronaldo said, "Manchester is in the past. Now my club is Real Madrid. This is my home." He also added, "In life there are things more important than money. It is important—I'm not going to lie—but the project is to win trophies. I feel integrated into this project." That was simply brilliant!

Let us construct the game. Suppose Real Madrid's valuation of Ronaldo's on the pitch performance is V per season, which can easily be in the range €20–€25 million. To start with, suppose Ronaldo's default payoff is €10 million per annum. Since he was already earning that, it can be safely argued that any of the top clubs in Europe would have agreed to pay him that much or more. In absence of any concrete alternative offer, the game tree looked like the one in Figure 8.12a.

In this game, knowing that Ronaldo will accept anything above €10 million, Real Madrid would have offered some x that is just above €10 million. With the offer from Manchester United, the game tree looked like the one in Figure 8.12b.

In presence of the offer from Manchester United, Real Madrid had no choice but to offer y larger than €16.67 million. Given that V is much more than €17 million, it makes business sense for Real Madrid to offer €17 million, and that is what they did. The strategic scenario is shown in Figure 8.13.

In terms of the value-net model, Ronaldo increased his bargaining power using the offer from Manchester United. A wage cap system, like in NBA or NFL in the USA, would effectively increase the bargaining power of the clubs. In NFL (the league of professional American football in the USA) the franchisees (clubs)

Figure 8.12a

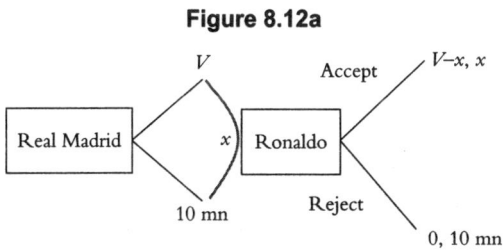

Figure 8.12b

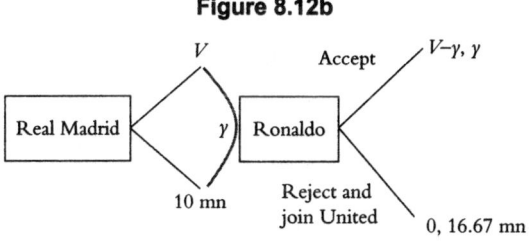

Figure 8.13

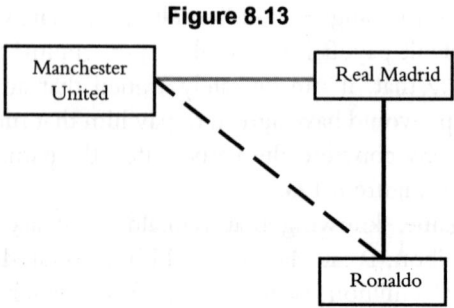

cannot spend more than the cap limit in total wages, which in turn acts as a commitment not to give exorbitant raise to even the most sought after players.

Making Commitment

Making commitment might help in gaining bargaining power. Recall our discussion on the MFC clause from section 'Leveraging Repeated Play to Out-think Customers' of Chapter 6. Signing a MFC clause with the buyers is effectively making a commitment not to reduce price. This commitment increases the bargaining power of the seller. In Chapter 6, we discussed the anti-trust case that was filed against Ethyl Corporation and DuPont, by the FTC of USA, for signing MFC clause with their respective buyers. New York Federal Court of Appeals, however, overturned the FTC ruling stating that Ethyl Corporation had MFC clauses with their customer even when they had no competition. In a competitive scenario the MFC clause acts as a commitment of not reducing price, made to the competitor. But even when there is no competition the MFC clause increases the minimum price that is acceptable to the seller, and thus increases the seller's bargaining power.

Consider a scenario where a seller is negotiating a deal with a new buyer. Suppose they play a two-stage alternate offer bargaining game with the seller first asking a price P_1. If that price is rejected by the buyer, the buyer will make a counter offer to

Figure 8.14a

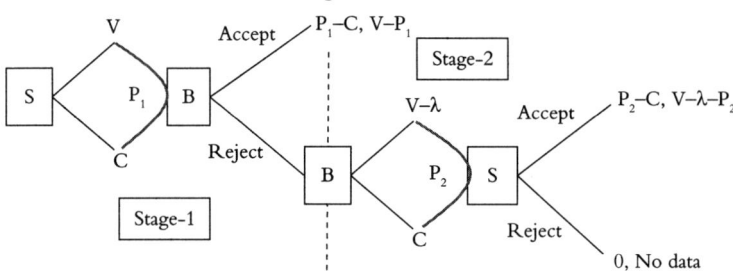

buy at the price P_2, which the seller decides to accept or reject. The game tree is given in Figure 8.14a. S is the seller and B is the buyer.

V is the buyer's valuation of the seller's product, and C is the seller's cost. The seller produces or procures only after the order is placed by the buyer, the seller does not incur the cost either. Hence, the default payoff to S is zero. We may not have information on B's default payoff. The buyer may or may not have an alternative to S's product. The buyer has a waiting cost of λ.

In stage-2, S will accept if $P_2 - C \geq 0$ or $P_2 \geq C$. So, in stage-2 B will offer $P_2 = C$. Foreseeing that in stage-2 the price will be C and their payoff will be $(V - \lambda - C)$, in stage-1 B will accept price P_1 if and only if $(V - P_1) \geq (V - \lambda - C)$, that is, $P_1 \leq \lambda + C$. So, in stage-1, S can at most ask for $P_1 = \lambda + C$.

Suppose S sold the same product earlier at a price P^* to another buyer, B^{old}, and signed a contract including an MFC clause with B^{old}. Now if S sells the product at any price $P < P^*$, the MFC clause makes him liable to pay $(P^* - P)$ to B^{old}. So, effectively the cost to S becomes $[C + (P^* - P)]$, and hence, while negotiating with the new buyer B, the minimum acceptable price to S becomes P^*. The game tree with this modification is given in Figure 8.14b.

In presence of an existing contract with an MFC clause, in stage-2 S will accept the price if $(P_2 - P^*) \geq 0$ or $P_2 \geq P^*$. Anticipating that, in stage-2 B will offer $P_2 = P^*$. Foreseeing that the stage-2 price will be P^* and their payoff will be $(V - \lambda - P^*)$; in stage-1 B will accept the price P_1 if and only if $(V - P_1) \geq$

Figure 8.14b

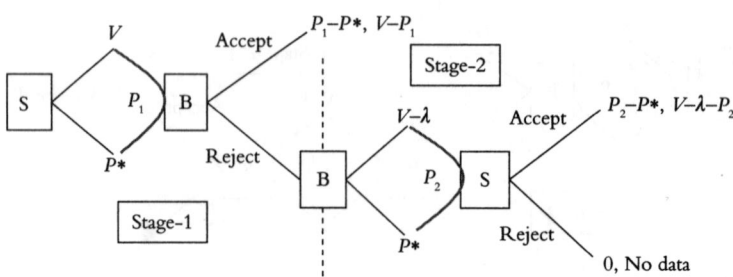

$(V - \lambda - P^*)$, that is, $P_1 \le \lambda + P^*$. So, in stage-1, S will ask for $P_1 = \lambda + P^*$. This example shows how MFC increases the bargaining power of the seller.

In this section, we discussed only a few tactical moves. There can be various other tactics to increase your bargaining power or to reduce the bargaining power of the opponent. But the fundamental idea remains the same. In order to increase your own bargaining power you need to increase your default payoff. That is possible by increasing your options. In order to increase your options you need to leverage the existence of other players in the value net of your business. As a buyer you would want to have more suppliers. As a seller you would want to have more buyers. Making commitments not to reduce price helps sellers. A commitment not to offer price above a cap helps the buyers. Also, collusion between sellers increases the seller's bargaining power. As discussed in Chapter 6, such collusion is only possible if the competing sellers perceive the game as an infinitely repeated game, and they don't heavily discount future. When the sellers collude using a trigger strategy, they know that their competitor will also not reduce the price. This confidence, in turn, increases their bargaining power vis-a-vis the buyers. Similarly, a buyer's collusion helps the buyers to keep price low.

Epilogue

I defined a game as a situation of strategic interaction between two or more individually rational players. The game may or may not be a game of pure conflict, where one player's gain is the others' losses. Such games of pure conflict are known as zero-sum or constant-sum games. In zero-sum games, the payoffs taken by the players sum up to a constant for all possible strategy profiles. The solution method of zero-sum games, known as maximin–minimax method, critically depends on this property of zero-sum games and predates John Nash. In this book, I purposely didn't distinguish between zero-sum games and non-zero-sum games. Instead of exposing the reader to too many concepts, I walked them through the essentials with the help of applications, exercises and cases. Maximin–minimax equilibrium is also Nash equilibrium and can be explained using the idea of best responses to each other. Interested readers may refer to Dixit and Skeath (2004). Avoiding any pre-Nash era equilibrium concept, I jumped straight to Nash equilibrium in Chapter 3 because it is a more general equilibrium concept and can be used in any simultaneous move game of complete information.

We understood the method of backward induction in Chapter 2. The equilibrium that we find using the method of backward induction is called sub-game perfect Nash equilibrium. I didn't use the technical definition of sub-game perfection in this book. As a matter of fact, I have not mentioned the term earlier. That was not an omission, but was done with purpose. Loosely speaking, sub-game perfection requires the actions of each player to be optimal in each node they move, on the equilibrium path as well as off it. I thought that my readers, whose primary purpose of learning game theory is to be able to use it in making decisions, can do without the complex technicalities of sub-game perfection. Understanding the idea of looking forward and reasoning backward is good enough for making decisions in sequential move games. It might help if one can draw the game tree. If we can draw a game tree and trace the decisions backward to find the equilibrium path connecting the end of the game tree to the starting node, what we find is a sub-game perfect Nash equilibrium. Any finer technicality beyond this might be superfluous for most of my readers. If they are interested, they may refer to Gibbons (1992) or Fudenberg and Tirole (1993) for a technical definition of sub-game perfect Nash equilibrium.

I have introduced my readers to the method of backward induction in Chapter 2 and to the basic concept of Nash equilibrium in Chapter 3. These two concepts were captured in two ideas. Backward induction was conceptualized by the idea of 'looking forward and reasoning backward' in sequential move games. The essential concept of Nash equilibrium was understood as strategies that are 'best responses to each other' in simultaneous move games. The rest of the book was built upon these two fundamental ideas. Chapter 4 continued with simultaneous move games, albeit without dominant strategies, and Chapter 5 introduced the strategic moves. Strategic moves are assets to a tactician. The concept was developed by Thomas Shelling and interested readers may refer to his famous book *The Strategy of Conflict.*

Repeated games are introduced in Chapter 6. Equilibrium concept used in repeated games is also sub-game perfect Nash equilibrium. I avoided getting technical and simply checked the conditions under which a player would deviate from the laid out trigger strategy. Compromising on the present value calculation was impossible, as the decision to cooperate or deviate depends on the present values. Mathematically inclined readers may refer to Myerson (1997) or Rubinstein and Osborne (1994) for a detailed exposition of folk theorem, or what Roger Myerson termed as general feasibility theorem.

Any kind of game that deals with probabilities was covered in Chapter 7. Mixed strategy Nash equilibrium could have been covered separately, or as part of Chapter 4. From my classroom experience I know that some people get a bit intimidated by probabilities. They can read the whole book bypassing Chapter 7. I defined mixed strategy loosely as a mix of pure strategies or a probability distribution over the set of pure strategies, and went on to find the best responses to each other in terms of mixed strategies. The equilibrium is called the mixed strategy Nash equilibrium. Actually, the mixed strategy Nash equilibrium is the generalized equilibrium conceptualized by John Nash, and the Nash equilibrium or equilibria in pure strategies is a subset of that generalized Nash equilibrium. Nash theorem says that there exists at least one Nash equilibrium in any finite game (games with finite number of players where each player has finite number of strategies), at least in mixed strategies. In this book, I never bothered with existence or stability of any equilibrium and didn't prove any theorem. Readers interested in the theorem and its proof may refer to Nash (1950, 1951). One needs to be conversant in mathematical analysis to understand the proof.

Chapter 7 introduced games of incomplete information with players having Bayesian types, and helped the readers to understand the concepts of 'type' and 'belief', and also how a player's prior belief can be represented by a random draw of 'nature'. With that understanding, I went on to solve sequential move games of incomplete information by the method of backward induction. Here, I avoided any further complexity. The equilibrium concept used for the solution of sequential move games of incomplete information is called sequential equilibrium. Later, the equilibrium criteria were refined and the more general concept of perfect Bayesian equilibrium was developed. For sequential move games of incomplete information with only two types of players, the perfect Bayesian equilibrium is the same as the sequential equilibrium. I didn't use these terms in Chapter 7. But the solution method I used for solving signalling games is essentially that of a perfect Bayesian equilibrium. To qualify as a perfect Bayesian equilibrium a strategy profile must satisfy the criteria of (a) belief consistency and (b) sequential rationality. In the solutions, I used both criteria without naming them and without getting too technical. If one wants to know more about sequential equilibrium and perfect Bayesian equilibrium, then refer to Kreps and Wilson (1982) and

Fudenberg and Tirole (1991). But it might be difficult to read these papers without proper training in advanced mathematics. Simultaneous move games of incomplete information were not covered in length. Instead, I went to an application of such games—auctions. Sealed-bid auctions are simultaneous move games of incomplete information. The equilibrium concept used for simultaneous move games of incomplete information is known as Bayesian Nash equilibrium. I bypassed the technical definition altogether. Instead, I solved an example of first price sealed-bid auction to get an idea of how decisions could be made. If one is interested, then he/she may read the three-article series of Harsanyi (1967–1968) for a technical definition of Bayesian Nash equilibrium. For the generalized results in auction and for the proof of revenue equivalence theorem refer to Krishna (2002).

Chapter 8 on negotiations is an out and out applied chapter. In it I used all the techniques that were introduced in the previous chapters to come up with a few rules for negotiations. Most game theory books discuss different kinds of bargaining games within the chapters dedicated for sequential move games under complete information and that under incomplete information. I took it out and created a separate chapter on negotiations to provide the readers with an idea of how we can bring different tools and techniques together to put them in use for decision-making in a particular kind of management problem. Ideally, auctions, pricing, contracting, etc., could have been put in separate chapters. But then, the initial chapters would have become all theory. I avoided that.

In this book, I didn't cover cooperative game theory, evolutionary game theory and behavioural game theory. In the Preface I mentioned why this book will not address evolutionary game theory and behavioural game theory. Cooperative game theory presumes that players can coordinate their strategies to arrive at a cooperative outcome. I believe that need for cooperation comes from self-interest and can be analysed within the framework of non-cooperative game theory. I discussed possibilities of collusion in the context of repeated games in Chapter 6. That is an example of how players might cooperate from an individually rational motive. Cooperative game theory discusses possibilities of coalition formation, existence of coalitions and stability of coalition at length, using equilibrium concept of 'core' and the 'Shapley value' criteria. Coalitional games are predominantly used in politics, but can be useful in assessing possibilities and stability of joint ventures. Due to limits in space and scope, I avoided it here. For a lucid and easy exposition to cooperative game theory, evolutionary game theory and behavioural game theory, I suggest the readers to go through the relevant chapters from McCain (2007). Shubik (1985, 1987) provides more rigorous exposition to cooperative game theory. For a comprehensive treatment of evolutionary game theory refer to Samuelson (1998). Behavioural and experimental game theory is covered well in Holt (2006).

Bibliography

Andersen, Steffen, Seda Ertaç, Uri Gneezy, Moshe Hoffman, and John A. List. 2011. "Stakes Matter in Ultimatum Games." *The American Economic Review*, 101 (7): 3427–39.

Arthur, W. B. 1996. "Increasing Returns and the New World of Business." *Harvard Business Review*, 72 (July): 100–109.

Brandenburger, Adam M. and Barry J. Nalebuff. 1997. *Co-opetition—A Revolutionary Mind-set that Combines Competition and Cooperation.* Crown Business.

Buchdahl, Ellie. 2013. "Tobacco Firms to Spend Millions on 'Glamorous' e-cigarette TV Commercials Nearly Two Decades After Advertising Ban." 14 September, *Mail Online*, http://www.dailymail.co.uk/health/article-2420693/Smoking-TV-Tobacco-firms-spend-record-amounts-e-cigarette-commercials-nearly-decades-advertising-ban.html (last accessed 28 July 2015).

Butler, Charlotte and Sumantra Ghoshal. 2002. "Hindustan Lever Limited: Levers for Change," in *World Class in India: A Casebook of Companies in Transformation*, eds, Sumantra Ghoshal, Gita Piramal and Sudeep Budhiraja. Penguin Books India.

Cooper Russell. 1998. *Coordination Games.* Cambridge University Press.

Dawkins, R. 2007. *The Selfish Gene.* Reprint. Oxford University Press.

Demsetz, H. 1982. "Barriers to Entry." *The American Economic Review*, 72 (1): 47–57.

Dixit, Avinash and Susan Skeath. 2004. *Games of Strategy.* W.W. Norton & Company.

Dutta, Prajit K. 1999. *Strategies and Games: Theory and Practice.* MIT Press.

Fudenberg, Drew and Jean Tirole. 1991. "Perfect Bayesian Equilibrium and Sequential Equilibrium." *Journal of Economic Theory*, 53 (2): 236–60.

———. 1993. *Game Theory.* MIT Press.

Ghemawat, P. 1997. *Games Businesses Play: Cases and Theory.* MIT Press.

Gibbons, Robert. 1992. *Game Theory for Applied Economists.* Princeton University Press.

Güth, Werner, Rolf Schmittberger and Bernd Schwarze. 1982. "An Experimental Analysis of Ultimatum Bargaining." *Journal of Economic Behavior and Organization*, 3 (4): 367–88.

Harsanyi, J. 1967. "Games with Incomplete Information Played by Bayesian Players: Part I. The Basic Model." *Management Science*, 14 (3), 159–82.

———. 1968a. "Games with Incomplete Information Played by Bayesian Players: Part II. Bayesian Equilibrium Points." *Management Science*, 14 (5): 320–34.

———. 1968b. "Games with Incomplete Information Played by Bayesian Players: Part III. The Basic Probability Distribution of the Game." *Management Science*, 14 (7): 486–502.

Holt, Charles A. 2006. *Markets, Games and Strategic Behavior—Recipes for Interactive Learning.* Pearson/Addison Wesley.

Kaiser, Ingrid, K. Jensen, J. Call and M. Tomasello. 2012. "Theft in an Ultimatum Game: Chimpanzees and Bonobos are Insensitive to Unfairness." *Biology Letters*, 8 (6): 942–45.

Katz, M. and C. Shapiro. 1994. "Systems Competition and Network Effects." *The Journal of Economic Perspectives*, 8 (2): 93–115.

Kreps, David M. and Robert Wilson. 1982. "Sequential Equilibria." *Econometrica*, 50 (4): 863–94.

Krishna, Vijay. 2002. *Auction Theory.* San Diego, CA: Academic Press.

Leyton-Brown, Kevin and Yoav Shoham. 2008. *Essentials of Game Theory: A Concise, Multidisciplinary Introduction.* Morgan & Claypool Publishers.

Mahdawi, Arwa. 2014. "Cigarette Advertising Back on TV: Decades of Campaigning up in Smoke." 18 February, *The Guardian*, http://www.theguardian.com/commentisfree/2014/feb/18/cigarette-advertising-decades-anti-smoking-campaigning (last accessed 28 July 2015).

McCain, Roger A. 2007. *Game Theory: A Nontechnical Introduction to the Analysis of Strategy.* World Scientific Publishing Company.

McNeilly, Mark R. 2001. *Sun Tzu and the Art of Modern Warfare.* Oxford University Press.

METRO Group. 2004. "RFID: Uncovering the Value." Metro AG (Düsseldorf).

Milgrom, P. and J. Roberts. 1982. "Limit Pricing and Entry under Incomplete Information." *Econometrica*, 50 (2): 443–60.

Milgrom, P. and R. J. Weber. 1982. "A Theory of Auctions and Competitive Bidding." *Econometrica*, 50 (5): 1089–1122.

Mitchell, Stacy. 2003. "German High Court Convicts Wal-Mart of Predatory Pricing," 1 February 2003, Independent Business, http://www.ilsr.org/german-high-court-convicts-walmart-predatory-pricing/ (last accessed 28th July 2015).

Morgenstern, Oskar and John von Neumann. 1947. *The Theory of Games and Economic Behavior.* Princeton University Press.

Myerson, Roger B. 1997. *Game Theory: Analysis of Conflict.* Harvard University Press.

Nash, John. 1950. "Equilibrium Points in n-person Games." *Proceedings of the National Academy of Sciences*, 36 (1): 48–49.

Nash, John. 1951. "Non-Cooperative Games." *The Annals of Mathematics*, 54 (2): 286–95.

Osborne, Martin. 2004. *An Introduction to Game Theory*. Oxford University Press.

Perarnau, Marti. 2014. *Pep Confidential*. Arena Sport.

Potts, Gregory. 2000. "Crest Sues Wal-Mart Over Edmond Store Pricing," 27 September 2000, NewsOK, http://newsok.com/crest-sues-wal-mart-over-edmond-stores-pricing/article/2713349 (last accessed 28th July 2015).

Poundstone, William. 1992. *Prisoner's Dilemma: John von Neumann, Game Theory, and the Puzzle of the Bomb*. Anchor Books, Doubleday.

Proctor, D. Williamson and Brosnan de Waal. 2013. "Chimpanzees Play the Ultimatum Game." *Proceedings of the National Academy of Sciences of the USA*, 110 (6): 2070–75.

Rapoport, Anatol and Albert M. Chammah. 1965. *Prisoner's Dilemma*. University of Michigan Press.

Ribeiro, John. 2015. "Uber's Indian rival Ola being probed for predatory pricing." 7 May, PC World.

Rubinstein, Ariel. 1982. "Perfect Equilibrium in a Bargaining Model." *Econometrica*, 50 (1): 97–109.

Rubinstein, Ariel and Martin J. Osborne. 1994. *A Course in Game Theory*. MIT Press.

Samuelson, Larry. 1998. *Evolutionary Games and Equilibrium Selection*. MIT Press

Schelling, Thomas C. 1966. *Arms and Influence*. New Haven, CN: Yale University Press.

———. 1980. *The Strategy of Conflict*. Reprint, Harvard University Press.

Selten, Reinhard. 1978. "The Chain Store Paradox." *Theory and Decision*, 9 (2): 127–59.

Shapiro, Carl and Hal R. Varian. 1999. *Information Rules*. Harvard Business School Press.

Shoham, Yoav and Kevin Leyton-Brown. 2009. *Multiagent Systems: Algorithmic, Game-Theoretic, and Logical Foundations*. Cambridge University Press.

Shubik, Martin. 1985. *Game Theory in Social Sciences* (Vol. 1)—*Concepts and Solutions*. MIT Press.

———. 1987. *Game Theory in Social Sciences* (Vol. 2)—*A Game Theoretic Approach to Political Economy*. MIT Press.

Shy, Oz. 1996. "Technology Revolutions in the Presence of Network Externalities." *International Journal of Industrial Organisation*, 14 (6): 785–800.

———. 2004. *The Economics of Network Industries*. Cambridge University Press.

Slonim, Robert and Alvin E. Roth. 1998. "Learning in High Stakes Ultimatum Games: An Experiment in the Slovak Republic." *Econometrica*, 66 (3): 569–96.

Smith, J. Maynard and G. R. Price. 1973. "Logic of Animal Conflict." *Nature*, 226 (02 November): 15–18.

Smith, J. Maynard. 1974. "Theory of Games and the Evolution of Animal Conflicts." *Journal of Theoretical Biology*, 47 (1): 209–21.

Spence, Michael. 1973. "Job Market Signaling." *The Quarterly Journal of Economics*, 87 (3): 355–74.

Trivers, R. L. 1971. "The Evolution of Reciprocal Altruism." *Quarterly Review of Biology*, 46 (1): 35–57.

Vickrey, William. 1961. "Counterspeculation, Auctions, and Competitive Sealed Tenders." *The Journal of Finance*, 16 (1): 8–37.

Wernerfelt, B. 1991. "Brand Loyalty and Market Equilibrium." *Marketing Science*, 10 (3): 229–45.

Wall Street Journal (Eastern edition). 2003. "Dogfight in the Secret World of Airplane Deals—One Battle up Close." 10 March, New York, A.1.

Stacy Mitchell. 2003. "German High Court Convicts Wal-Mart of Predatory Pricing", 1 February 2003, Independent Business http://www.ilsr.org/german-high-court-convicts-walmart-predatory-pricing/ (last accessed 28th July 2015).

Gregory Potts. 2000. "Crest Sues Wal-Mart Over Edmond Store Pricing." 27 September 2000, NewsOK, http://newsok.com/crest-sues-wal-mart-over-edmond-stores-pricing/article/2713349 (last accessed 28th July 2015).

Index

Acme Wagon Co. *vs* Selco Steel Inc.,
case of alternate bargaining, 179–84
added-value, 166
aggressive strategy, 80–83
Airbus, 28–30, 37–39, 62–63, 75–76
airlines industry, 28–30
Alan's War-The Memories of G.I. Alan Cope (Emmanuel Guibert), 5
alternate offer bargaining game, 177–84
alternate offer wage bargaining, 186–92
altruism, 3
anti-coordination games, 48, 70. *See also* coordination games
chicken game, 61–62
hawk-dove game, 60–61
mixed strategy in, 123–26
assurance games, 40–47
payoff matrix, 41, 47
auctions, 157–63, 204
Dutch, 159–60
English, 158–59
first price sealed-bid auction, 160–61
revenue equivalence, 163
Vickery, 161–63

backward induction, 18, 202
bargaining power, tactics to improve, 194–200
battle of sexes (BoS), 49–51, 54
Bayesian Nash equilibrium, 204
'best alternative to negotiated agreement' (BATNA), 173

bluffing, 115, 131
Boeing, 28–30, 37–38, 62–63, 75–76
Bowling, Michael, 115
bridge game, 144–45
British Satellite Broadcasting (BSB), 66–67
buyer–seller negotiations, 165, 177

Café Coffee Day (CCD), 9, 18–19, 21
cartelization, 107
certainty equivalent of the lottery, 122
chess game, 10–14
chicken game, 61–62. *See also* commitment
game between Airbus and Boeing, 62–63
Chomp game, 14–18
Christmas Truce of 1914, 93
cigarette/tobacco advertisement, ban on, 43–46
coalitional games, 205
collective wage bargaining, 184–94, 204
alternate offer wage bargaining, 186–92
fundamentals of, 194
strikes and, 192–94
tactical issues in negotiating, 194–200
commitment
chicken game, role of, 61–63
by eliminating options, 74–80
gaining credibility through, 72–74

computing system, Macintosh *vs* Windows, 53
cooperative game theory, 204–5
Co-opetition (Adam Brandenburger and Barry Nalebuff), 6
coordination games, 48, 70. *See also* anti-coordination games
battle of sexes (BoS), 49–51
problem of technology adoption and network effect, 52–60
pure, 51–52
cordial relationship and cooperation, 92–93
Costa Coffee, 2, 64, 86–88
Cristal USA Inc., 108–9

dictator game, 168–70
discounting factor, 99
discounts offer, 37–40
dominated strategy, 21, 116
strategic moves using, 80–83
DuPont Co., 108–9
Dutch auction, 159–60

English auction, 159–60
Entrant *vs* Incumbent, illustration of a generic game, 19–22
expected payoffs, 119–20

FC Barcelona *vs* Real Madrid, 2009, 5
first mover's advantage, 27–30
first price sealed-bid auction, 160–61
focal equilibrium, 42
four-stage alternate offer bargaining game, 180

Gainesville Regional Utilities (GRU) *vs* CSX, 6–7
Gale, David, 14
game, description of a, 3–4
games of incomplete information, 130–35
auctions as, 157–63
player types and belief, 131–35

game theory
with abstract payoffs, 51
first mover's advantage, 27–30
payoffs of a player, 4, 22–27
understanding, 3–4
use of, 4, 8
game tree, 23–26, 71–75, 78–79, 81, 84–85, 133, 140, 142
grim trigger strategy (GTS), 95–101
Guardiola, Pep, 4–5

hawk–dove game, 60–61
'heads up or tails up', game of, 116–18, 120–21
higher education as a signal of ability, 151–57
Hindustan Lever Ltd (HLL) *vs* Nirma, 82–83

Iberia, 37–39
incentives, 83–85
Radio Frequency Identification (RFID) technology, in retail supply chain, 83–85
individual rationality, 3
individual risk behaviour, 121–23

Kronos, 108–9

lock-in customers, 64
looking forward and reasoning backward, idea of, 14–18

market expansion game, 63–69
mathematical expectation, 119–20
maximin–minimax equilibrium, 201
mixed strategies
in anti-coordination games, 123–26
applications, 126–29
mixed strategy Nash equilibrium, 118–19, 202–3
in one-shot game, 120–21
most favoured customer (MFC) clause, 109–12

mutually assured destruction (MAD)
strategy, 95
mutual preening, 91–92

Nash equilibrium, 35–36, 40–42, 46,
54, 68–69, 80–81, 83, 96, 116–17,
125, 129, 201
in coordination games, 46, 50
mixed strategy, 118–21
in pure strategies, 127
sub-game perfect, 201–2
negative reciprocity, 169–71
negotiations
negotiations, tactical issues in,
194–200
network effect, 52–60
problem of technology adoption,
52–53

Ola taxi hailing service in India, 143
one-shot game, 95–96
reputation, importance of, 135
organizational culture, influence in
game play, 1
Organization of the Petroleum
Exporting Countries (OPEC),
trigger strategies of, 103–8
cartelization, 107
price fixing by implicit collusion,
108
out-thinking phenomenon
by credibility, 6–7
by doing the unimaginable, 5–6
by exploiting the weakness of rival,
4–5

passive strategy, 80–82
payoffs, 4, 39, 49, 51, 54, 59–60, 62,
78, 80, 83, 123–26
matrix for the one-shot game, 96
understanding, case example, 22–27
Pep Confidential (Marti Perarnau), 4
pirate ship problem, 174–77. *See also*
ultimatum game

players of a game, 3
poker game, 115–16
"Texas Hold'em" version of, 116
pre-commitment, 77
predatory pricing, 136–44
as a signal of cost, 145–50
pre-emptive retreat, 76–77
present discounted value (PDV), 7,
99–101
price fixing by implicit collusion,
108–9
prisoner's dilemma game, 32–34, 89
in matrix form, 33–34
modified. *See* assurance games
Nash equilibrium of, 36
option of discounts, 37–40
in terms of strictly dominant
strategy, 34–35
pure coordination games, 51–52
pure strategies, 123, 127

Radio Frequency Identification
(RFID) technology, in retail supply
chain, 54–60, 83–85
benefits, 59
at Metro AG, 56–58
payoffs, 58–59
reciprocal altruism, 93
reciprocal relationship, 93–94
repeated games, 90–94, 202
mutual preening, case of 'grudger'
birds, 91–92
reciprocal altruism, 93
reciprocal relationship, 93–94
trigger strategies and, 94–107
reputation, 135–36
risk-averse bidders, 161
risk-lovers, 122–23
risk-neutral bidders, 161
risk-neutral individuals, 123
risk premium, 122

Schuh, Frederik, 14
sealed-bid auctions, 157, 204

self-esteem of an individual, 173
self-interested behaviour, 3–4
Selfish Gene (Richard Dawkins), 91
sequential move game, 27–29, 31, 62,
  78, 203
  Entrant *vs* Incumbent, illustration
    of a generic game, 19–22
  Starbucks *vs* CCD, 18–19
signalling, 144–57
simultaneous move games, 30–31,
  202, 204
  payoff matrix form for, 31–34
stability of collusion, conditions for
  number of players, 102–3
  reaction lag of the players, 102
  total capacity of firms, 103
Starbucks, 2, 9
  business strategy, 9, 18–19, 21
  in China, 64, 86–88
  entry in India, 9, 18–19
strategic moves
  by changing order, 70–72
  gaining credibility through
    commitment, 72–74
  with incentives, 83–85
  making commitment by
    eliminating options, 74–80
  strategic location choice, 86–88
  use of dominant strategy, 80–83
strategic thinking, 2
strategies, 4
strategy profile, 35, 50
strictly dominant strategies, 34–35
sub-game perfect Nash equilibrium,
  201–2

Tammelin, Oskari, 116
16th Armoured Division *vs* Nazis,
  Second World War, 5–6

tit-for-tat (TfT) strategy, 101
tossing a coin, 116–18
Toys"R"Us, 113–14
trigger strategies, 92, 94–107
  cartelization, 107
  conditions for stability of collusion,
    102–3
  fundamental principles, 94–95
  grim trigger strategy (GTS), 95–101
  lessons from theory and history,
    101–3
  meet the competition clause
    (MCC), 112–14
  most favoured customer (MFC)
    clause, 109–12
  of Organization of the Petroleum
    Exporting Countries (OPEC),
    103–7
Trivers, R. L., 93
two-stage alternate offer bargaining
  game, 177–78

UEFA Champions' League *vs* Vienna
  Philharmonic Orchestra, choosing
  between, 49–51
ultimatum game, 167–74
  dictator game, 168–69
  experimental findings, 168
  with high stakes, 171–73
  negative reciprocity, 169–71

value-net model, 164–66
Vickery auctions, 161–63

Walmart, 136–38
war of attrition, 63–69, 128–29
  in satellite television market of UK,
    65–67
weakly dominant strategy, 41

About the Author

Dr Sumit Sarkar is presently Professor of Economics at XLRI, Jamshedpur. He is a PhD from the Centre for Economic Studies and Planning, JNU, New Delhi. Before joining XLRI, he was a member of the faculty at IIT Kanpur and at IIM Kozhikode. He has also taught at some other IIMs as visiting faculty. His area of research is Applied Game Theory, Industrial Organization Theory, Economics of Network Industries and Auction Theory. His research papers were published in various international journals including *The Manchester School, International Journal of Management and Network Economics, Information Technology and People*, etc. Dr Sarkar is also a reputed trainer. He conducts management development programmes for middle and senior management. In the past, he has conducted training sessions or modules for managers of Forbes Marshall Pvt. Ltd, Tata Chemicals Ltd, Mahindra Finance, Mashreq Bank (UAE), NHPC Ltd, Hindustan Aeronautics Ltd, Oil India Ltd, Indian Ordinance Factories, etc.